String Analysis for Software Verification and Security

Tevfik Bultan • Fang Yu • Muath Alkhalaf
Abdulbaki Aydin

String Analysis for Software Verification and Security

 Springer

Tevfik Bultan
Department of Computer Science
University of California, Santa Barbara
Santa Barbara, CA, USA

Muath Alkhalaf
Computer Science Department
King Saud University
Riyadh, Saudi Arabia

Fang Yu
Department of Management
Information Systems
National Chenchi University
Taipei, Taiwan

Abdulbaki Aydin
Microsoft (United States)
Redmond, WA, USA

ISBN 978-3-319-88637-4   ISBN 978-3-319-68670-7   (eBook)
https://doi.org/10.1007/978-3-319-68670-7

Printed on acid-free paper

This Springer imprint is published by Springer Nature
The registered company is Springer International Publishing AG
The registered company address is: Gewerbestrasse 11, 6330 Cham, Switzerland

Preface

This monograph is mainly based on the research that has been conducted in the Verification Laboratory at the University of California, Santa Barbara, in the last decade. String analysis has been an interesting and fruitful area to work on, leading to many research results some of which are discussed here. We observe that the research in analysis of string manipulating code is expanding due to the importance of the correctness of the string manipulation code for dependability and security of modern software systems. We hope that this monograph can inspire and motivate more research in this area and accelerate the transition of string analysis research to practice.

We would like to thank all current and past members of the Verification Laboratory for their help and support. We would also like to acknowledge the support provided by the National Science Foundation (NSF) and the Defense Advanced Research Projects Agency (DARPA) for the string analysis research at the Verification Laboratory.[1]

Santa Barbara, CA, USA	Tevfik Bultan
Taipei, Taiwan	Fang Yu
Riyadh, Saudi Arabia	Muath Alkhalaf
Redmond, WA, USA	Abdulbaki Aydin

[1] NSF under grants CCF 0916112, CNS-1116967, CCF-1423623, and CCF-1548848, and DARPA under agreement number FA8750-15-2-0087. The U.S. Government is authorized to reproduce and distribute reprints for Governmental purposes notwithstanding any copyright notation thereon. The views and conclusions contained herein are those of the authors and should not be interpreted as necessarily representing the official policies or endorsements, either expressed or implied, of DARPA or the U.S. Government.

Contents

Chapter 1
Introduction

In computing, a sequence of characters is called a *string*. For example, "Hello, World!" is a string (we use double quotes to indicate the beginning and end of the string). This particular string is very familiar to programmers since the first programming assignment in many programming textbooks is to write a program that outputs the string "Hello, World!" Although programmers become familiar with the creation and manipulation of strings very early on in their training, errors in string manipulating code is a major cause of software faults and vulnerabilities. This indicates that string manipulation is a challenging task for programmers, and automated techniques for analyzing string manipulating code, which is the topic of this monograph, are very desirable.

Handling of strings in programming languages vary widely. In the C programming language, characters are a basic data type, but strings are not. Strings are represented as arrays of characters, which is a natural way to store strings. Since strings do not have a type dedicated to them, basic operations on strings (such as concatenating two strings) are not part of C, and are instead provided as library functions (such as `strcat`, `strcpy`, `strcmp`, `strlen`).

In the Java programming language, strings are objects corresponding to the instances of `String` class. String operations are provided as the methods of the `String` class (such as the `length`, `charat` and `concat` methods) and the `String` class is implemented by storing strings as arrays of characters. Java also provides a string concatenation operator ("+").

In the more recent JavaScript programming language, strings are supported as one of the primitive data types, and strings can be constructed using operators such as string concatenation operator "+". JavaScript also provides a set of methods for string manipulation (such as `substring`, `indexOf`, `replace`).

We see that different languages treat strings differently. In general, support for string manipulation in programming languages has been increasing. This is due to increasing use of string manipulation in implementation of modern software

© Springer International Publishing AG 2017
T. Bultan et al., *String Analysis for Software Verification and Security*,
https://doi.org/10.1007/978-3-319-68670-7_1

applications. Here are some common uses of string manipulation in modern software development:

- *Input sanitization and validation:* Many modern software applications are web-based and the user inputs to web applications typically come in the string form (for example a text field entered by the user). Since any user with an Internet connection can access web applications and interact with them, web application developers have to assume that the user input could be malicious. There are well-known cyber-attack techniques that involve a malicious user submitting hidden commands via user input fields. Execution of these harmful commands can lead to unauthorized access to or loss of data. To prevent such scenarios, application developers have to either reject the user input that does not fit expected patterns (which is called input validation), or clean up the user input by removing unwanted characters (which is called input sanitization). Both input validation and sanitization involve string manipulation since user inputs are typically in string form.
- *Query generation for back-end databases:* Many modern software applications use a back-end database to store data. When users interact with modern software applications, user requests result in generation of a query that is sent to a back-end database. The user request triggers a piece of code that constructs the database query as a string. So, string manipulation is an essential part of query generation.
- *Generation of data formats such as XML, JSON, and HTML:* Modern software applications typically use well-known data formats to store, exchange, or describe data. XML and JSON are two of the widely used data formats. HTML is the format used to describe web documents and to display user input forms in web applications. Many modern applications dynamically create documents in XML, JSON, or HTML format during program execution. Creation of XML, JSON, or HTML documents involve manipulation of strings.
- *Dynamic code generation:* Software applications are becoming increasingly dynamic. For example, many modern web applications dynamically generate code based on user requests. Web applications are multi-tiered systems where part of the code is executed on web servers (that are typically hosted in compute-clouds), and part of the code is executed on the client machine. It is common for the server-side code to generate client-side code dynamically at runtime. Client-code is generated using string manipulation.
- *Dynamic class loading and method invocation:* Programming languages are also becoming increasingly dynamic. Modern languages such as JavaScript and PHP allow functions to be specified using string variables, where the invoked methods or loaded classes depend on the values of the string variables at runtime. Reflection in Java allows programs to load classes dynamically at runtime. Similarly, in Objective-C, one can load classes from string variables. This provides developers powerful means to adjust program executions according to their runtime environment and status. However, since the loaded classes or invoked methods depend on the values of string variables at runtime, malicious

developers could manipulate string values to obfuscate the program behavior, and prevent static detection of malicious behaviors such as accessing sensitive/private APIs in Android/iOS mobile applications.

Due to extended use of strings in modern software development, errors in string manipulating code or maliciously written string manipulating code can have disastrous effects. It would be very helpful for developers to be able to automatically check if the string manipulation code works correctly with respect to their expectations. This is the core problem that *string analysis* addresses. String analysis is a static program analysis technique that determines the values that a string expression can take during program execution at a given program point. String analysis can be used to solve many problems in modern software systems that relate to string manipulation, such as:

- String analysis can be used to identify security vulnerabilities by checking if a security sensitive function can receive an input string that contains an exploit [78, 119, 130, 132]. This type of vulnerabilities are common in web applications when user input is not adequately validated or sanitized.
- Modern dynamic languages enable execution of dynamically generated code. For example, in JavaScript, the `eval` function can be used to execute dynamically generate code. The `eval` function takes a string value as an argument and executes the JavaScript expression, variable, statement, or sequence of statements that is given as the argument. So, in order to analyze the behavior of a JavaScript program that uses the `eval` function, we first need to understand the set of string values that can be passed to the `eval` function as an argument. String analysis can be used for this purpose [57].
- For applications that dynamically generate data in XML, JSON or HTML format, string analysis can be used to identify formatting errors in the generated documents, which would then identify bugs in the XML, JSON or HTML generating code [78].
- For applications that dynamically generate queries for a back-end database, string analysis can be used to identify the set of queries that are sent to the back-end database by analyzing the code that generates the SQL queries. This can be used to identify vulnerabilities, or to generate test queries for the back-end database [118, 119].
- For web applications where server-side code dynamically generates client-side code, string analysis can be used to determine the client-side code that can be generated [82, 83], and this can be used to analyze and find potential problems in the generated client-side code.
- String analysis can also be used for automatically repairing faulty string manipulation code. For example, input validation and sanitization functions can be repaired automatically by identifying the set of input values that can cause a vulnerability [128].
- For programs that have classes loaded from strings, string analysis can be used in advance to find all potential values of loaded classes or invoked methods, for example, for taming reflection in Java programs [22, 72]. Particularly, string

analysis can be used to analyze complex string manipulation to improve the precision of static taint/flow analysis.

Like many other program-analysis problems, it is not possible to solve the string analysis problem precisely (i.e., it is not possible to precisely determine the set of string values that can reach a program point). However, one can compute over- or under-approximations of possible string values. If the approximations are precise enough, they can enable us to demonstrate existence or absence of bugs and vulnerabilities in string manipulating code. String analysis has been an active research area in the last decade, resulting in a wide variety of string-analysis techniques such as, grammar-based string analysis [63, 78], automata-based symbolic string analysis [26, 34, 55, 113, 114], string constraint solving [3, 12, 15, 71, 93, 112], string abstractions [21, 131], relational string analysis [133], vulnerability detection using string analysis [100, 119, 129], differential string analysis [5, 7], and automated repair using string analysis [5, 128].

1.1 Common Vulnerabilities Due to String Manipulation Errors

Let us discuss the security vulnerabilities related to string manipulation errors further. Security vulnerabilities in web applications are common. Ubiquity and global accessibility of web applications make this a critical problem. Malicious users all around the world can exploit vulnerable web applications resulting in unauthorized access to private data or loss of critical information. Since web applications are continuously in use without any available downtime for manual analysis or repair, vulnerabilities in web applications should not only be discovered fast, but should also be repaired fast. String analysis techniques can be used to address this problem.

The Common Vulnerabilities and Exposures (CVE) repository keeps track of all reported computer security vulnerabilities [30]. According to the CVE repository, web application vulnerabilities form a significant portion of all reported computer security vulnerabilities. Two of the most common vulnerabilities in web applications are: Cross-site Scripting (XSS) and SQL Injection (SQLI).

A XSS vulnerability results from the application inserting part of a malicious user's input in an HTML page that it renders. For example, a malicious user can attack a web forum application by posting a comment that contains a link that contains an executable script (such as JavaScript code) within the text of the comment. The attack occurs, if, while viewing the comment, another user clicks on the link that contains the malicious code. The victim's browser will then execute the malicious JavaScript code and this can result in stealing of browser cookies and other sensitive data.

An SQL Injection vulnerability, on the other hand, results from the application's use of user input in constructing database queries. The attacker can invoke the application with a malicious input that becomes part of an SQL command that the

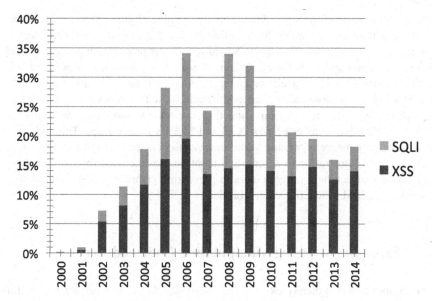

Fig. 1.1 Percentage of the Cross-site Scripting (XSS) and SQL Injection (SQLI) vulnerabilities among all the computer security vulnerabilities reported in the CVE repository [30]

application executes. This permits the attacker to damage or get unauthorized access to data stored in the back-end database of the application.

Figure 1.1 shows what percentage of all reported computer security vulnerabilities are of these two types over the years, based on the vulnerabilities reported in the CVE repository. Additionally, Open Web Application Security Project (OWASP) compiles a top ten list to identify the most critical security flaws in web applications [84–86]. According to the OWASP top ten lists compiled in 2007, 2010 and 2013, XSS and SQLI are always among the top three web application vulnerabilities.

XSS and SQLI vulnerabilities are caused by inadequate input validation and sanitization. Any error in validation and sanitization of user input can lead to a significant vulnerability for a web application, and can be exploited by malicious users all around the world. Input validation and sanitization in web applications is crucial and challenging since communication between different layers of a web application occurs through directives (commands) that often embed user input and are written in many languages, such as XML, SQL, and HTML. Programs that propagate and use malicious user inputs without validation and sanitization, or with inadequate validation and sanitization, are vulnerable to attacks such as XSS and SQLI.

One thing to note in Fig. 1.1 is that, the percentage of XSS and SQLI vulnerabilities among all reported security vulnerabilities increased significantly between 2000 and 2006. This can be attributed to the significant increase in web applications during that time period, and the lack of techniques and tools and lack of awareness among software developers for eliminating these vulnerabilities. We see a decrease in these vulnerabilities after 2006, although the number of web applications continued to increase. This can be explained with better programming practices and better techniques and tools for eliminating these vulnerabilities. However, we still see that more than 15% of all computer security vulnerabilities are due to XSS and SQLI flaws in more recent years. Although, these two types of vulnerabilities in web applications are notorious, and they are widely publicized, we continue to see a significant number of new XSS and SQLI vulnerabilities. This indicates the challenging nature of writing string manipulation code for implementing input validation and sanitization functions, and demonstrates the need for static analysis techniques that can help programmers in preventing errors in string manipulating code that can lead to security vulnerabilities.

1.2 Examples of String Manipulating Code and Errors

A web application typically expects the user input to be in a certain format for many text fields (such as username, email, zip code, etc.). Since the user input can contain typing errors, or may be purposefully written (by a malicious user) to violate the expected format, the web application has to validate the user input using input validation operations such as regular expression matching. Furthermore, the web application may need to modify the input to put it in the expected format using sanitization operations such as trimming white spaces from the beginning and the end of an input or escaping problematic characters. Figure 1.2 shows an example of a validation and sanitization function written in Java to validate and sanitize email addresses in a web application called JGossip (http://sourceforge.net/projects/

```
1   public class Validator {
2     public boolean validateEmail(Object bean, Field f, ..) {
3       String val = ValidatorUtils.getValueAsString(bean, f);
4       Perl5Util u = new Perl5Util();
5       if (!(val == null || val.trim().length == 0)) {
6         if ((!u.match("/( )|(@.*@)|(@\\.)/", val))
7           && u.match("/^[\\w]+@([\\w]+\\.[\\w]{2,4})$/",
8                       val)) {
9           return true;
10        } else {
11          return false;
12        }
13      }
14      return true;
15    }
16    ...
17  }
```

Fig. 1.2 Server-side input validation code snippet written in Java

```
function validate() {
  ...
  switch(type) {
   case "time":
    var highlight = true;
    var default_msg = "Please enter a valid time.";
    time_pattern = /^[1-9]\:[0-5][0-9]\s*(\AM|PM|am|pm?)\s*$/;
    time_pattern2 = /^[1-1][0-2]\:[0-5][0-9]\s*(\AM|PM|am|pm?)\s*$/;
    time_pattern3 = /^[1-1][0-2]\:[0-5][0-9]\:[0-5][0-9]\s*(\AM|PM|
                      am|pm?)\s*$/;
    time_pattern4 = /^[1-9]\:[0-5][0-9]\:[0-5][0-9]\s*(\AM|PM|
                      am|pm?)\s*$/;
    if (field.value != "") {
     if (!time_pattern.test(field.value)
        && !time_pattern2.test(field.value)
        && !time_pattern3.test(field.value)
        && !time_pattern4.test(field.value)) {
          error = true;
        }
    }
    break;
   case "email":
    error = isEmailInvalid(field);
    var highlight = true;
    var default_msg = "Please enter a valid email address.";
    break;
   case "date":
    var highlight = true;
    var default_msg = "Please enter a valid date.";
    date_pattern = /^(\d{1}|\d{2})\/(\d{1}|\d{2})\/(\d{2}|\d{4})\s*$/;
    if (field.value != "")
     if (!date_pattern.test(field.value)||!isDateValid(field.value))
       error = true;
    break;
  ...
}
```

Fig. 1.3 Part 1 of a client-side JavaScript code example demonstrating the complexity of input validation and sanitization code

jgossipforum/). This is a server-side validation function that is executed after the user input reaches to the web server. Line 5 checks that the email address value is not null or empty after trimming space characters. Lines 6 and 7 validate the input by matching against a regular expression using the Perl5Util package's match function. Note that regular expressions used in this type of validation code can be quite complex and prone to errors.

Input validation code can be more complex than the example shown in Fig. 1.2. Consider the example code shown in Figs. 1.3 and 1.4. This example shows parts of a typical input validation and sanitization code used to validate a form in a Google website. This is client-side code written in JavaScript. It is executed at the user's machine. This type of client-side input validation is commonly used since checking the user input before sending it to the server can improve the responsiveness of the

```
function validate() {
  ...
  if (alert_msg == "" || alert_msg == null) alert_msg = default_msg;
  if (error) {
   any_error = true;
   total_msg = total_msg + alert_msg + "|";
  }
  if (error && highlight) {
   field.setAttribute("class","error");
   field.setAttribute("className","error");      // For IE
  }
  ...
}
```

Fig. 1.4 Part 2 of a client-side JavaScript code example demonstrating the complexity of input validation and sanitization code

```
1 <?php
2     $www = $_GET["www"];
3     $1_otherinfo = "URL";
4     $www = preg_replace( "/[^A-Za-z0-9 .-@://]/", "", $www );
5     echo $1_otherinfo . ": " . $www;
6 ?>
```

Fig. 1.5 A server-side input sanitization code snippet written in PHP

application by preventing unnecessary communication over the network, and it can also reduce the load of the server-side machines. Note that, this code (1) mixes the input validation and sanitization of multiple HTML form fields at the same time, and (2) mixes the actual code that does the input validation and sanitization with other parts of the program that do error reporting and event handling. Due to its complexity, it would not be surprising to encounter errors in this type of code, and it is difficult to analyze it.

Let us now look at some examples of erroneous input validation and sanitization code to see what kind of bugs appear in real-world code. Figure 1.5 shows a simplified version of a vulnerable PHP sanitization code that is taken from a web application called MyEasyMarket [13].

The code starts with assigning the user input provided in the $_GET array to the $www variable in line 2. Then, in line 3, it assigns a string constant to the $1_otherinfo variable. Next, in line 4, the user input is sanitized using the preg_replace command. This replace command gets three arguments: the match pattern, the replace string and the target string. It finds all the substrings of the target string that match the match pattern and replaces them with the replace string. In the replace command shown in line 4, the match pattern is the regular expression [^A-Za-z0-9 .-@://], the replace string is the empty string (which corresponds to deleting all the substrings that match the match pattern), and the target string is the value of the variable $www. After the sanitization step, the PHP code outputs the concatenation of the variable $1_otherinfo, the string constant ": ", and the variable $www.

The replace operation in line 4 contains an error that leads to a XSS vulnerability. The error is in the match pattern of the replace operation: [^A-Za-z0-9 .-@://]. The goal of the programmer was to eliminate all the characters that should not appear in a URL. The programmer implements this by deleting all the characters that do not match the characters in the regular expression [A-Za-z0-9 .-@://]. The programmer's intention is to eliminate everything other than alpha-numeric characters, and the ASCII symbols ., -, @, :, and /. However, the regular expression is not correct. First, there is a harmless error. The subexpression // can be replaced with / since repeating the symbol / twice is unnecessary. More serious error is the following: The expression . -@ is the union of all the ASCII symbols that are between the symbol . and the symbol @ in the ASCII ordering. The programmer intended to specify the union of the symbols ., -, and @ but forgot that symbol - has a special meaning in regular expressions when it is enclosed with symbols [and]. The correct expression should have been . \-@. This error leads to a vulnerability because the symbol < (which can be used to start a script to launch a XSS attack) falls between the symbol . and the symbol @ in the ASCII ordering. So, the sanitization operation fails to delete the < symbol from the input, leading to a XSS vulnerability.

Using string replace operations, such as the one in this example, to sanitize user input is common practice in web applications. However, this type of sanitization is error prone due to complex syntax and semantics of regular expressions.

Figure 1.6 shows a JavaScript email validation function taken from a telecommunication company website (www.stc.co.sa). This function has a different problem than the previous one. The previous PHP function was under constrained and accepts bad inputs while this function is over constrained and rejects good inputs. In line 10, the input email address is validated against the complex regular expression in lines 5–9. The problem is that this regular expression does not allow email addresses with capital letters. Although this problem is present in the client-side of

```
1   function isValidEmail(emailField) {
2          var email = emailField.value.trim();
3          emailField.value = email;
5          EMAIL_REGEXP =
6          /^[a-z0-9!#$%&'*+/=?^_`{|}~-]+
7          (?:\.[a-z0-9!#$%&'*+/=?^_`{|}~-]+)*@
8          (?:[a-z0-9](?:[a-z0-9-]*[a-z0-9])?\.)+
9          [a-z0-9](?:[a-z0-9-]*[a-z0-9])$/;
10     if(!EMAIL_REGEXP.test(email)) {
11          alert("Please enter a correct email address.");
12          emailField.focus();
13          return false;
14     }
15     return true;
16 }
```

Fig. 1.6 A client-side email validation code snippet written in JavaScript

the web application, it will affect the application's correctness since it will prevent some valid emails with capital letters to reach the server.

Web application developers often introduce redundant input validation and sanitization code in the client and server-side code of a web application. The input provided by the user can be validated and/or sanitized by the client-side code (written usually in JavaScript) that is executed on the user's machine to make sure it is in the correct format. As we mentioned earlier, the advantage of validating user input on the client-side (instead of doing it exclusively on the server-side) is that it improves usability and responsiveness of the application by preventing unnecessary communication with the server and reduces the server load at the same time. When the user input reaches the server, the server-side code validates and/or sanitizes the user input again. Although this may sound redundant, it is necessary due to the fact that the users can be malicious, and a malicious user can bypass the client-side validation by manually crafting the HTML request with malicious input. So, it is necessary to validate and/or sanitize the user input again at the server-side. However, this introduces possibility of another type of bug. The validation and sanitization policies at the client and server-side code could be inconsistent.

Figure 1.7 shows the client-side input validation function in JGossip web application (written in JavaScript) that corresponds to the server-side input validation function (written in Java) in Fig. 1.2. Although both functions validate the same

```
1  <html>
2  ...
3  <script>
4  function validateEmail(form) {
5    var emailStr = form["email"].value;
6    if(emailStr.length == 0) {
7      return true;
8    }
9    var r1 = new RegExp("( )|(@.*@)|(@\\.)");
10   var r2 = new RegExp("^[\\w]+@([\\w]+\\.[\\w]{2,4})$");
11   if(!r1.test(emailStr) && r2.test(emailStr)) {
12     return true;
13   }
14   return false;
15 }
16 </script>
17 ...
18 <form name="subscribeForm" action="/Unsubscribe"
19     onsubmit="return validateEmail(this);">
20   Email: <input type="text" name="email" size="64" />
21   <input type="submit" value="Unsubscribe" />
22 </form>
23 ...
24 </html>
```

Fig. 1.7 Client-side input validation code snippet written in JavaScript that corresponds to the server-side code in Fig. 1.2 written in Java

```
1 NSBundle *b = [NSBundle bundleWithPath:@"/System/Library
                /Frameworks/AdSupport.framework"];
2 if(b){
3  NSString *name = [NSString stringWithFormat:@"%s%s%s",
         "AS","Identifier","Manager"];
5  Class c = NSClassFromString(name);
6  id si = [c valueForKey:@"sharedManager"];
7 }
```

Fig. 1.8 Dynamic class loading with objective C in iOS applications

HTML input field that is used to input an email address, they return different results for the same input. On one hand, the client-side validation function rejects a sequence of one or more white space characters, for which the condition on line 6 evaluates to false and the regular expression check on line 11 fails, thereby resulting in the function returning false. However, for the same input, the second condition on line 5 of the server-side validation function (Fig. 1.2) evaluates to false, due to the trim function call, and the string is therefore accepted by the server. Accepting white spaces as email addresses by the server might lead to failures (for example by inserting invalid data to the back-end database). As a general policy, the server-side input validation/sanitization code should be more strict than the client-side input validation/sanitization code. If an application violates this policy, it could lead to erroneous behavior or security vulnerabilities.

In iOS programming, Objective-C allows programmers to load classes from strings. In contrast to static compilation, dynamic class loading provides runtime benefits such as function adjustment and delay loading until the code is needed. However, during dynamic class loading, developers may bypass static checking of potential exploits of iOS private/sensitive API usages. Figure 1.8 shows an example that loads class ASIdentifierManager dynamically. The class enables the developers to derive a shared instance of ASIdentifierManager to access users' advertisement data. After loading the framework AdSupport at line 1, the class ASIdentifierManager is dynamically loaded from a string variable name by calling the C-function NSClassFromString at line 3. Note that the value of name is composed using three strings "AS", "Identifier", and "Manager". That is to say, until the runtime, the loaded class is unknown and the class name ASIdentifierManager does not even appear in the code. Once the class is loaded, it then gets the value of a static field named sharedManager at line 6. Use of more complex string manipulation operations, such as replacement, or using an external input value in construction of the class name can obfuscate the loaded class further. Identifying NSClassFromString as a sensitive function, string analysis can be used to derive its input values to uncover the loaded class statically.

Similarly, Java Reflection makes it possible to inspect classes, interfaces, fields and methods at runtime, without knowing the names of the classes, methods etc. at compile time. This enables Java developers instantiate new objects, invoke methods and get/set field values using reflection in applications (including Android mobile applications). Figure 1.9 shows the fragment of an Android benchmark app

```
1 TelephonyManager telephonyManager = (TelephonyManager)
    getSystemService(Context.TELEPHONY_SERVICE);
2 String imei = telephonyManager.getDeviceId(); //source
3 Class c = Class.forName("de.ecspride.ReflectiveClass");
4 Object o = c.newInstance();
5 Method m = c.getMethod("setIme" + "i", String.class);
6 m.invoke(o, imei);
```

Fig. 1.9 A reflective call example from an Android app (from DroidBench [37])

that uses reflective call to access the device ID. The reflective class is loaded at line 3 and the method composed by two strings "setIme" and "i" at line 5 is invoked to set the device ID. In this case, string analysis can be used in advance to discover that the method "setImei" (a sensitive method) has been invoked.

1.3 Overview

Below we provide a brief overview of the contents of the remaining chapters:

- In Chap. 2 we first present a simple language for writing programs with string variables, and demonstrate that analyzing string manipulating programs is as hard as analyzing programs in general. We then present an extended string manipulation language to illustrate the types of string manipulating operations that are commonly used in software development.
- In Chap. 3 we first provide a transition-system semantics for string manipulating programs. We then discuss reachability analysis techniques starting with explicit state exploration. We present fixpoint characterization of reachability analysis using pre- and post-condition functions.
- In Chap. 4 we discuss the basic techniques for automata-based symbolic reachability analysis of string manipulating programs. We discuss the symbolic automata representation based on Multi-Terminal Binary Decision Diagrams (MTBDD) and present the forward and backward symbolic reachability analysis algorithms. Finally, we discuss how to compute the pre- and post-condition functions for basic string manipulation operations using automata as a symbolic representation.
- In Chap. 5 we discuss relational string analysis using multi-track automata as a symbolic representation. Relational string analysis is necessary for verification of properties that involve relations among multiple string variables.
- In Chap. 6 we discuss several automated abstraction techniques for string manipulating programs. We also present an automata-based widening operation for approximating fixpoints in symbolic reachability analysis of string manipulating programs.

- In Chap. 7 we discuss constraint-based string analysis. We present an automata-based approach to string constraint solving. We end this chapter with a discussion on model counting for string constraints.
- In Chap. 8 we discuss how string analysis techniques can be used for automatically detecting security vulnerabilities in web applications. We also present techniques for automatically repairing identified vulnerabilities.
- In Chap. 9 we discuss differential string analysis techniques. We start with a formal characterization of input validation and sanitization functions. We then present techniques for discovering inconsistencies between client- and server-side input validation and sanitization functions in web applications. We also present differential repair techniques for eliminating identified inconsistencies.
- In Chap. 10 we present a set of analysis tools which implement the techniques discussed in earlier chapters:

  - LIBSTRANGER is an automata-based symbolic analysis library that supports pre- and post-condition computations for string operations and widening operation.
  - STRANGER is an automata-based static vulnerability analysis tool for checking input validation and sanitization vulnerabilities in PHP applications.
  - SEMREP is a language agnostic tool for automatically repairing vulnerabilities in web applications using differential analysis.
  - ABC is an automata-based, model counting, string constraint solver.

- In Chap. 11 we present a brief survey of related research results.
- In Chap. 12 we provide our closing comments.

Chapter 2
String Manipulating Programs and Difficulty of Their Analysis

The goal of string analysis is to determine the values that a string expression can take during program execution at a given program point. Like many program analysis problems, this turns out to be a difficult problem. We can demonstrate the difficulty of string analysis using a simple language for string manipulation.

2.1 A Simple String Manipulation Language

We define the syntax of our simple string manipulation language in Fig. 2.1. Each program consists of a sequence of labeled statements. Statements can be assignment statements, conditional or unconditional branch statements, input and output statements, and assert statements. We have only one string operation (concatenation ".") and only one predicate (equality "=") in this simple language.

In Fig. 2.2 we show a simple program in the simple string manipulation language we described above. This program consists of two read statements followed by two assignment statements that use string concatenation followed by a conditional branch statement that uses the equality predicate followed by print and halt statements.

Assume that we want to check if an assertion in a program written in this language can ever fail. If we can do precise string analysis on programs written in this language, then we can determine all possible values for all string expressions in the program. So, then, we can check if it is possible to have an assertion violation. Note that, we can easily transform assertion violation checks to reachability checks where we check if a program statement is reachable. Given an assertion statement

```
assert exp;
```

© Springer International Publishing AG 2017
T. Bultan et al., *String Analysis for Software Verification and Security*,
https://doi.org/10.1007/978-3-319-68670-7_2

$$prog \rightarrow (lstmt)^+$$
$$lstmt \rightarrow l : stmt$$
$$stmt \rightarrow v := sexp;$$
```
       | if bexp then goto l;
       | goto l;
       | read v;
       | print sexp;
       | assert bexp;
       | halt;
```
$$bexp \rightarrow v = sexp \mid bexp \wedge bexp \mid bexp \vee bexp \mid \neg bexp$$
$$sexp \rightarrow v \mid "c" \mid sexp.sexp$$

Fig. 2.1 Grammar for a very simple string manipulation language

Fig. 2.2 A simple string
manipulating program

```
1: read x1;
2: read x2;
3: x1 := x1 . "a";
4: x2 := x2 . "a";
3: if (x1 = x2) goto 7;
5: print x1 . x2;
6: halt;
7: print x1;
```

we can easily transform it to a reachability check as follows:

```
   if exp goto 2;
1: print "assertion violation";
2: print "assertion holds";
```

Checking the reachability of the statement 1 is equivalent to checking if it is possible to have an assertion violation during some execution of the program.

2.2 Automated and Precise Verification of String Programs Is Not Possible

Let us define the *halting problem* for string programs as the problem of deciding, given a string program P, where initially the string variables are initialized to the null string, ϵ, whether P will halt (i.e., reach the halt statement) on some execution. More generally, the *reachability problem for string programs* (which need not halt) is the problem of deciding, given a string program P and a program state s (where a program state s is defined with the instruction label of an instruction in the program and the values of all the variables), whether at some point during a computation, the program state s will be reached. Note that, we define the halting and the reachability conditions using existential quantification over the execution paths, i.e., the halting and the reachability conditions hold if there exists an execution path that halts or reaches the target state, respectively. Hence, if the halting problem is undecidable, then the reachability problem is undecidable.

It can be shown that the halting program for string programs is undecidable [134]. In fact, halting problem for string programs is undecidable even if we restrict the number of string variables to 3. It is possible to show this by demonstrating that string programs can simulate counter machines.

Counter machines are a simple and powerful computational model that can simulate Turing Machines. A counter machine consists of a finite number of counters (unbounded integer variables) and a finite set of instructions. Counter machines have a very small instruction set that includes an increment instruction, a decrement instruction, a conditional branch instruction that tests if a counter value is equal to zero, and a halt instruction. So, during the execution of a counter machine, at each step, a counter can be incremented by 1, decremented by 1, and tested for zero. The counters can only assume nonnegative values. It is well-known that the halting problem for two-counter machines, where both counters are initialized to 0, is undecidable [79]. In fact, two counter machines can simulate Turing Machines.

We show that a two-counter machine M can be simulated by a string program P with only three string variables X_1, X_2, X_3. The program counter of M can be represented as labels in the string program P. The instructions where the counter-machine M halts will be represented with the halt instruction in the string program P. We will use the lengths of the strings X_1, X_2 and X_3 to simulate the values of the counters C_1 and C_2. The value of C_1 will be simulated by $|X_1| - |X_3|$, and the value of C_2 will be simulated by $|X_2| - |X_3|$.

The counter machine M starts from the initial configuration $(q_0, 0, 0)$ where q_0 denotes the initial instruction and the two integer values represent the initial values of counters C_1 and C_2, respectively. The initial state of the string program P will be $(q_0, \epsilon, \epsilon, \epsilon)$ where q_0 is the label of the first instruction, and the strings $\epsilon, \epsilon, \epsilon$ are the initial values of the string variables X_1, X_2 and X_3, respectively. The instructions of the counter-machine C can be simulated by instructions of a string program as shown in Fig. 2.3 where each statement is followed by a goto statement that transitions to the next instruction.

Counter machine instruction	String program simulation
inc C_1	$X_1 := X_1.a;$
inc C_2	$X_2 := X_2.a;$
dec C_1	$X_2 := X_2.a; \; X_3 := X_3.a;$
dec C_2	$X_1 := X_1.a; \; X_3 := X_3.a;$
if $(C_1 = 0)$	if $(X_1 = X_3)$
if $(C_2 = 0)$	if $(X_2 = X_3)$
halt	halt;

Fig. 2.3 Simulation of instructions of a counter machine with string program instructions

Note that although this transformation allows the simulated counter values to possibly take negative values, this can be fixed by adding a conditional branch instruction before each decrement that checks that the simulated counter value is not zero before the instructions simulating the decrement instruction is executed. The string program P constructed from the counter machine M based on these rules will simulate M. Hence, solving the halting problem for string programs would mean that we can solve the halting problem for counter machines. Since halting problem for counter machines is undecidable, we conclude that halting problem for string programs is also undecidable. Since 2-counter machines can simulate Turing Machines we also observe that string programs with only 3 string variables can simulate Turing Machines.

This discussion demonstrates that automatically and precisely checking reachability properties of string manipulating programs is not possible. Reachability problem for string manipulating programs in undecidable, hence it cannot be precisely solved like many other program analysis problems. However, it is possible to check reachability properties of string manipulating programs approximately. For example, it would be possible to develop automated and approximate techniques that guarantee there are no assertion violations (i.e., that there a no bugs) for some programs. However, such *sound* verification techniques could sometimes report *false positives*, i.e., report assertion violations although no execution of the program may have assertion violations. Similarly, it would be possible to develop automated and approximate techniques that guarantee that there are assertion violations in some programs. However, such *complete* verification techniques could sometimes report *false negatives*, i.e., they may not find an assertion violation even though in some execution of the program there may be an assertion violation.

When a program analysis problem is undecidable, it means that there does not exist a sound and complete technique to solve it. Since we showed that reachability for string programs is an undecidable problem, we conclude that there is no automated verification technique for string programs that can report assertion violations for all string programs that contain assertion violations, and that can report the absence of assertion violations for all string programs that do not contain assertion violations.

2.3   A Richer String Manipulation Language

We have seen that, even with 3 string variables and a small instruction set, string programs can simulate Turing Machines. Adding more string operations does not increase the expressiveness of string manipulating programs theoretically. However, in practice, string analysis has to deal with many other string manipulation operations in addition to string concatenation. In Fig. 2.4 we give the syntax for an extended string manipulation language.

Let us briefly describe the semantics of extended string operations below. We assume that Σ denotes the set of all characters (i.e., the alphabet) in our string manipulation language, and Σ^* denotes the set of all strings (including the

Fig. 2.4 Grammar for an
extended string manipulation
language

$$
\begin{aligned}
prog &\rightarrow block \\
block &\rightarrow lstmt^+ \\
lstmt &\rightarrow l \;:\; stmt \\
stmt &\rightarrow v := exp; \\
&\mid \; \texttt{read}\, v; \\
&\mid \; \texttt{print}\; exp; \\
&\mid \; \texttt{assert}\; bexp; \\
&\mid \; \texttt{halt}; \\
&\mid \; \texttt{if}\,(bexp)\,\texttt{then}\,\{block\} \\
&\mid \; \texttt{if}\,(bexp)\,\texttt{then}\,\{block\}\,\texttt{else}\,\{block\} \\
&\mid \; \texttt{while}\,(bexp)\,\{block\} \\
exp &\rightarrow sexp \mid iexp \\
bexp &\rightarrow sexp = sexp \\
&\mid \; \texttt{match}(sexp, sexp) \\
&\mid \; \texttt{contains}(sexp, sexp) \\
&\mid \; \texttt{begins}(sexp, sexp) \\
&\mid \; \texttt{ends}(sexp, sexp) \\
&\mid \; iexp = iexp \mid iexp < iexp \mid iexp > iexp \\
&\mid \; bexp \wedge bexp \mid bexp \vee bexp \mid \neg bexp \\
iexp &\rightarrow v \mid n \mid iexp + iexp \mid iexp - iexp \\
&\mid \; \texttt{length}(sexp) \\
&\mid \; \texttt{indexof}(sexp, sexp) \\
sexp &\rightarrow v \mid ``c'' \mid sexp.sexp \mid sexp^* \mid sexp \mid sexp \\
&\mid \; \texttt{replace}(sexp, sexp, sexp) \\
&\mid \; \texttt{substring}(sexp, iexp, iexp) \\
&\mid \; \texttt{charat}(sexp, iexp) \\
&\mid \; \texttt{reverse}(sexp)
\end{aligned}
$$

empty string ϵ). The operations ".", "$*$", and "$\mid$" correspond to regular expression operations concatenation, Kleene closure and alternation.

- $\texttt{match}(s, r)$ means that string s matches the string expression r (which could be a regular expression). Let $\mathcal{L}(r)$ denote the set of strings (i.e., the language) defined by the string expression r, then we define the semantics of $\texttt{match}$ as:

$$
\texttt{match}(s, r) \Leftrightarrow s \in \mathcal{L}(r)
$$

- $\texttt{contains}(s, t)$ means that string t is a substring of string s:

$$
\texttt{contains}(s, t) \Leftrightarrow \exists s_1, s_2 \in \Sigma^* : s = s_1 t s_2
$$

- $\texttt{begins}(s, t)$ means that string s begins with string t:

$$
\texttt{begins}(s, t) \Leftrightarrow \exists s_1 \in \Sigma^* : s = t s_1
$$

- ends(s, t) means that string s ends with string t:

$$\text{ends}(s, t) \Leftrightarrow \exists s_1 \in \Sigma^* : s = s_1 t$$

- length(s) denotes the length of the string s (for empty string, length $(\epsilon) = 0$):

$$(\text{length}(s) = 0 \Leftrightarrow s = \epsilon) \wedge (\text{length}(s) = n \Leftrightarrow \exists c_1, c_2, \ldots, c_n \in \Sigma : s = c_1 c_2 \ldots c_n)$$

We also use $|s|$ to denote the length of string s, i.e., $\text{length}(s) = |s|$.

- indexof(s, t) denotes the smallest starting location of substring t in string s. If x is not a substring of s (i.e, $\neg \text{contains}(s, t)$) then $\text{indexof}(s, t) = -1$.

$$(\text{indexof}(s, t) = -1 \Leftrightarrow \neg \text{contains}(s, t)) \wedge$$
$$(\text{indexof}(s, t) = n \quad \Leftrightarrow \quad (\exists s_1, s_2 \in \Sigma^* : s = s_1 t s_2 \wedge |s_1| = n)$$
$$\wedge (\forall i < n : \neg(\exists s_1, s_2 \in \Sigma^* : s = s_1 t s_2 \wedge |s_1| = i)))$$

- replace(s, p, t) replaces the pattern string p in s with the target string t and returns the result.

$$r = \text{replace}(s, p, t) \Leftrightarrow ((\neg \text{contains}(s, p) \wedge r = s) \vee$$
$$(\exists s_3, s_4, s_5 \in \Sigma^* : s = s_3 p s_4 \wedge r = s_3 t s_5 \wedge s_5 = \text{replace}(s_4, p, t) \wedge$$
$$(\forall s_6, s_7 \in \Sigma^* : s = s_6 p s_7 \Rightarrow |s_6| \geq |s_3|)))$$

Note that, this description of the replace function semantics assumes that pattern string p is a string value. In general, p could be a regular expression and more than one substring of the source string s can match the regular expression p. In that case, the semantics has to specify which matching substring will be chosen for replacement. Typically, the longest matching substring is chosen for replacement.

- substring(s, i, j) returns the substring of string s that starts at index i (inclusive) and ends at index j (exclusive).

$$t = \text{substring}(s, i, j) \Leftrightarrow \exists s_1, s_2 \in \Sigma^* : s = s_1 t s_2 \wedge |s_1| = i \wedge |t| = j - i$$

- charat(s, i) returns the character that appears at the index i of the string s. The semantics of charat is defined as follows:

$$t = \text{charat}(s, i) \Leftrightarrow \exists c_0, c_1, \ldots, c_n \in \Sigma : s = c_0 c_1 \ldots c_n \wedge 0 \leq i \leq n \wedge t = c_i$$

- reverse(s) returns the reverse of the string v.

$$t = \text{reverse}(s) \Leftrightarrow \exists c_0, c_1, \ldots, c_n \in \Sigma : s = c_0 c_1 \ldots c_n \wedge t = c_n \ldots c_1 c_0$$

```php
1  <?php
2     $www = $_GET["www"];
3     $l_otherinfo = "URL";
4     $www = preg_replace( "/[^A-Za-z0-9 .-@://]/", "", $www );
5     echo $l_otherinfo . ": " . $www;
6  ?>
```

(a)

```
2:  read www;
3:  l_otherinfo := "U" . "R" . "L";
4:  www = replace( ..., "", www);
5:  print l_otherinfo . ":" . " " . www;
```

(b)

Fig. 2.5 A code snippet (**a**) written in PHP and the corresponding code snippet (**b**) in the string manipulation language

It is possible to extend the string manipulation language defined in Fig. 2.4 even further by adding more variations of string operations we defined. For example, the replace operation we defined replaces all occurrences of substrings that match the pattern. A variant of the replace operation replaces only the first appearance of the substring that matches the given pattern. Yet another variant replaces only the last appearance of the substring that matches the given pattern. As another example, the indexof operation we defined returns the index of the first appearance of the given substring. A variant of this operation is the last-index-of operation that returns the index of the last appearance of the given substring. The techniques we discuss in this monograph can be extended to handle such extensions, but to keep our discussions concise we restrict our scope to the operations defined in Fig. 2.4.

Semantics of string manipulating programs can be defined based on the semantics of string manipulation operations we gave above.

Consider the PHP program segment that we showed in Chap. 1, Fig. 1.5. We can model this function using the string manipulation language we introduced as shown in Fig. 2.5. Note that the lines that correspond to each other are labeled with the same line number. Since we did not introduce a complex regular expression syntax in our string manipulation language, we did not provide the translation for the regular expression used in the line 4 of the PHP code snippet. However, it is possible to translate this PHP regular expression to a simple regular expression by just taking disjunction of all the characters that correspond to the PHP regular expression.

String analysis techniques developed for the string manipulation language we introduced in this chapter can be adopted to string analysis problems in modern programming languages by either extracting a string program from a given program by identifying the string expressions, or by implementing the string analysis techniques in a static analysis tool for the given programming language.

2.4 Summary

In this chapter we first presented a basic set of string manipulating instructions and then provided a larger set of instructions with more complex string operations. We discussed the difficulty of analyzing string manipulating programs. If we assume that string variables can take arbitrarily large string values, then even with a basic set of instructions, verification of string manipulating programs is an undecidable problem. Hence, verification of string manipulating programs cannot be fully automated and precise at the same time. In the following chapters we discuss automated techniques that use approximations and abstractions for string analysis.

Chapter 3
State Space Exploration

Before we can discuss string analysis and verification techniques for string manipulating programs, we first need to discuss the semantics of string manipulating programs in more detail.

3.1 Semantics of String Manipulation Languages

Semantics of string manipulating programs we defined in Chap. 2 can be formalized as transitions systems. A transition system $T = (I, S, R)$ consists of a set of states S, a set of initial states $I \subseteq S$ and a transition relation $R \subseteq S \times S$. Let us first define the set of states of a string manipulating program written either in the language shown in Fig. 2.1 or the extended language shown in Fig. 2.4. Each program state will correspond to a program location. If we assume that all statements are labeled with unique labels, we can use the labels of the statements to denote the possible program locations. Let us use L to denote the set of program locations (i.e., the set of statement labels). Each string variable will have a value from the set Σ^* and each integer variable will have a value from the set $\mathbb{Z}$. Let us assume that we are given a program with n string and m integer variables. Then, the set of states of the given program is defined as:

$$S = L \times (\Sigma^*)^n \times (\mathbb{Z})^m$$

Let us assume that $l_1 \in L$ denotes the label of the first statement of the program. Then we can define the set of initial states of the program as:

$$I = \{\langle l_1, \epsilon, \ldots, \epsilon, 0, \ldots, 0 \rangle\}$$

© Springer International Publishing AG 2017
T. Bultan et al., *String Analysis for Software Verification and Security*,
https://doi.org/10.1007/978-3-319-68670-7_3

where the program counter is initialized to l_1, all string variables are initialized to ϵ (i.e., empty string) and all integer variables are initialized to 0. Since we are assuming all the variables are initialized, and that there is a single initial statement, the set of initial states is a singleton set. It would not be a singleton set if we assume that the variables are not initialized and their initial values are not known.

The transition relation (R) of the program is defined by the semantics of the statements of the program. The semantics of the statements can be inferred from the semantics of the string operations we defined in Chap. 2.

Given a statement labeled l, let $r_l \subseteq S \times S$ denote tuples of program states $(s_1, s_2) \in r_l$ such that, s_1 is a program state at the program location l, and executing statement l in program state s_1 results in the program state s_2. So, r_l denotes the transition relation of the statement l. Then the transition relation of the whole program can be defined as:

$$R = \bigcup_{l \in L} r_l$$

i.e., taking the union of all the transitions defined by each statement of the program gives us the transition relation of the whole program.

Note that, the transition relation of a string manipulating program can be an infinite state system if we do not put a bound on the lengths of the strings. The undecidability of the reachability problem for string programs that we discussed in Chap. 2 is due to unboundedness of the string variables. In our formal model for string manipulating programs, each string variable has an infinite set of possible values (which is the set Σ^*), and hence, set of states of a string manipulating program is infinite.

Since no computer has an infinite amount of memory, bounding all the domains is a practical approach to program analysis and verification. However, when a program is analyzed or verified based on a given bound, obtained results are not guaranteed to hold when the program execution exceeds that bound. So, assuming an infinite state space is a useful assumption if one wants to check a program's behavior for arbitrarily large state spaces.

Let us consider the string program example shown in Fig. 3.1. This program has three string variables x, y and z. The initial state of this program is: $\langle l, x, y, z \rangle = \langle 1, \epsilon, \epsilon, \epsilon \rangle$ which means that initially program counter is 1, and string variables x, y, z are initialized to the empty string ϵ.

The transition relation of the program is defined as the union of the transition relations of its instructions:

$$R = r_1 \cup r_2 \cup r_3$$

Fig. 3.1 A simple string manipulating program with three string variables

```
1: x := "ab";
2: y := "cd";
3: z := x . y;
```

Fig. 3.2 A string
manipulating program that
copies a read string value
character by character

```
1: read x;
2: while (i < length(x)) {
3:     y := y . charat(x, i);
4:     i := i + 1;
5: }
```

and

$$r_1 = \{((\langle 1, \epsilon, \epsilon, \epsilon \rangle, \langle 2, ab, \epsilon, \epsilon \rangle))\}$$
$$r_2 = \{((\langle 2, ab, \epsilon, \epsilon \rangle, \langle 3, ab, cd, \epsilon \rangle))\}$$
$$r_3 = \{((\langle 3, ab, cd, \epsilon \rangle, \langle 4, ab, cd, abcd \rangle))\}$$

where we assume that 4 is an implicit halt instruction.

Let us consider the string program example shown in Fig. 3.2. This program reads a string value to the variable x and then it copies the value of string variable x to string variable y one character at a time. The initial state of this program is $\langle l, x, y, i \rangle = \langle 1, \epsilon, \epsilon, 0 \rangle$ which means that initially program counter is 1, string variables x and y are initialized to the empty string ϵ, and the integer variable i is initialized to 0.

The transition relation of the program is defined as the union of the transition relations of its instructions:

$$R = r_1 \cup r_2 \cup r_3 \cup r_4 \cup r_5$$

We assume that a read instruction can read any possible string value. So, the transition relation for the instruction 1, r_1 consists of an infinite set of transitions, where for each $s \in \Sigma^*$:

$$(\langle 1, \epsilon, \epsilon, 0 \rangle, \langle 2, s, \epsilon, 0 \rangle) \in r_1$$

Let us consider a state $\langle 2, ab, \epsilon, 0 \rangle$ where the value read to variable x is the string "ab". Following transitions are in the transition relation of this string program:

$$(\langle 2, ab, \epsilon, 0 \rangle, \langle 3, ab, \epsilon, 0 \rangle) \in r_2$$
$$(\langle 3, ab, \epsilon, 0 \rangle, \langle 4, ab, a, 0 \rangle) \in r_3$$
$$(\langle 4, ab, a, 0 \rangle, \langle 2, ab, a, 1 \rangle) \in r_4$$
$$(\langle 2, ab, a, 1 \rangle, \langle 3, ab, a, 1 \rangle) \in r_2$$
$$(\langle 3, ab, a, 1 \rangle, \langle 4, ab, ab, 1 \rangle) \in r_3$$
$$(\langle 4, ab, ab, 1 \rangle, \langle 2, ab, ab, 2 \rangle) \in r_4$$
$$(\langle 2, ab, ab, 2 \rangle, \langle 5, ab, ab, 2 \rangle) \in r_2$$

where we assume that label 5 corresponds to the termination of the loop.

3.2 Explicit State Space Exploration

When the semantics of a program is defined as a transition system (S, I, R), assertion checking corresponds to checking reachability in this transition system.

Let us consider the statements of the program. Each statement l has corresponding transition relation r_l. Using r_l we can also define a POST $: S \rightarrow S$ function as follows:

$$s_2 = \text{POST}(s_1, l) \Leftrightarrow (s_1, s_2) \in r_l$$

POST(s_1, l) denotes the state that the program can go by executing statement l at program state s_1. Note that, in the above definition, we are assuming that the transition system is deterministic, i.e., each state has at most one state that can be reached from it after one step of execution. We can generalize to nondeterministic systems if we allow POST function to return a set of states rather than a single state (as we discuss in the next section).

We can also define the POST function for the overall program as follows:

$$s_2 = \text{POST}(s_1) \Leftrightarrow \exists l \in L \; : \; s_2 = \text{POST}(s_1, r_l)$$
$$s_2 = \text{POST}(s_1) \Leftrightarrow (s_1, s_2) \in R$$

We can think of the POST function as computing the post-condition (or post-image) of a given state.

For the string program example shown in Fig. 3.1, we have the following:

$$
\begin{aligned}
\text{POST}((1, \epsilon, \epsilon, \epsilon), 1) &= \text{POST}((1, \epsilon, \epsilon, \epsilon)) &= (2, ab, \epsilon, \epsilon) \\
\text{POST}((2, ab, \epsilon, \epsilon), 2) &= \text{POST}((2, ab, \epsilon, \epsilon)) &= (3, ab, cd, \epsilon) \\
\text{POST}((3, ab, cd, \epsilon), 3) &= \text{POST}((3, ab, cd, \epsilon)) &= (4, ab, cd, abcd)
\end{aligned}
$$

Similarly, for the string program example shown in Fig. 3.2, we have the following:

$$
\begin{aligned}
\text{POST}((2, ab, \epsilon, 0)) &= (3, ab, \epsilon, 0) \\
\text{POST}((3, ab, \epsilon, 0)) &= (4, ab, a, 0) \\
\text{POST}((4, ab, a, 0)) &= (2, ab, a, 1) \\
\text{POST}((2, ab, a, 1)) &= (3, ab, a, 1) \\
\text{POST}((3, ab, a, 1)) &= (4, ab, ab, 1) \\
\text{POST}((4, ab, ab, 1)) &= (2, ab, ab, 2) \\
\text{POST}((2, ab, ab, 2)) &= (5, ab, ab, 2)
\end{aligned}
$$

Algorithm 1 REACHABILITYDFS

 1: $Stack := I$;
 2: $RS := I$;
 3: **while** $Stack \neq \emptyset$ **do**
 4:     $s :=$ POP$(Stack)$;
 5:     $s' :=$ POST(s);
 6:     **if** $s' \notin RS$ **then**
 7:         $RS := RS \cup \{s'\}$;
 8:         PUSH$(Stack, s')$;
 9:     **end if**
10: **end while**
11: **return** RS;

3.2.1 Forward Reachability

Let $RS(I)$, or simply RS, denote the set of states that are reachable from the initial states I of a program, i.e.,

$$RS = \{s \mid \exists s_0, s_1, \ldots, s_n : \forall i < n : (s_i, s_{i+1}) \in R \wedge s_0 \in I \wedge s_n = s\}$$

Using the POST function we can write a simple depth first search algorithm for computing reachable states of a program as shown in Algorithm 1.

For the string program example shown in Fig. 3.1, the set of reachable states RS can be computed using the algorithm shown in Algorithm 1, and the result would be:

$$RS = \{\langle 1, \epsilon, \epsilon, \epsilon \rangle, \langle 2, ab, \epsilon, \epsilon \rangle, \langle 3, ab, cd, \epsilon \rangle, \langle 4, ab, cd, abcd \rangle)\}$$

For the program shown in Fig. 3.2, the reachable states can be characterized as follows:

$$\langle l, s_1, s_2, i \rangle \in RS \Leftrightarrow \begin{array}{l} l = 1 \wedge s_1 = \epsilon \wedge s_2 = \epsilon \wedge i = 0 \\ \vee\ l = 2 \wedge s_1, s_3 \in \Sigma^* \wedge s_1 = s_2.s_3 \wedge i = \texttt{length}(s_2) \\ \vee\ l = 3 \wedge s_1, s_3 \in \Sigma^* \wedge s_1 = s_2.s_3 \wedge i = \texttt{length}(s_2) \\ \vee\ l = 4 \wedge s_1, s_3 \in \Sigma^* \wedge s_1 = s_2.s_3 \wedge i = \texttt{length}(s_2) - 1 \\ \vee\ l = 5 \wedge s_1 \in \Sigma^* \wedge s_1 = s_2 \wedge i = \texttt{length}(s_2) \end{array}$$

The set of states S and the set of reachable states RS for the program shown in Fig. 3.2 are infinite. For infinite states spaces, the explicit state exploration approach shown in Algorithm 1 would not terminate, so we need to find a different approach. However, before we address this issue, let us consider backward reachability problem.

3.2.2 Backward Reachability

Similar to the POST function we can also define a PRE function for backward reachability. Even for deterministic systems one state can have multiple states that can reach it in one step, so we need to define the PRE : $S \to 2^S$ function as follows:

$$s_2 \in \text{PRE}(s_1, l) \Leftrightarrow (s_2, s_1) \in r_l$$
$$s_2 \in \text{PRE}(s_1) \quad \Leftrightarrow \exists l \in L : s_2 \in \text{PRE}(s_1, r_l)$$
$$s_2 \in \text{PRE}(s_1) \quad \Leftrightarrow (s_2, s_1) \in R$$

We can think of the PRE function as computing the pre-condition (or pre-image) of a state.

For the string program example shown in Fig. 3.1, we have the following:

$$\text{PRE}(\langle 2, ab, \epsilon, \epsilon \rangle) \quad = \{\langle 1, \epsilon, \epsilon, \epsilon \rangle\}$$
$$\text{PRE}(\langle 3, ab, cd, \epsilon \rangle) \quad = \{\langle 2, ab, \epsilon, \epsilon \rangle\}$$
$$\text{PRE}(\langle 4, ab, cd, abcd \rangle) = \{\langle 3, ab, cd, \epsilon \rangle\}$$

We can find all states that can reach a particular target state using a depth first search algorithm similar to the one shown in Algorithm 1 that starts from the target state and uses the PRE function to compute backward reachability as shown in Algorithm 2.

Using the Algorithm 2 we can compute the backward reachability set for a given set of states. For the string program example shown in Fig. 3.1, we can compute the following sets:

$$BRS(\langle 3, ab, cd, \epsilon \rangle) \quad = \{\langle 1, \epsilon, \epsilon, \epsilon \rangle, \langle 2, ab, \epsilon, \epsilon \rangle\}$$
$$BRS(\langle 4, ab, cd, abcd \rangle) = \{\langle 1, \epsilon, \epsilon, \epsilon \rangle, \langle 2, ab, \epsilon, \epsilon \rangle, \langle 3, ab, cd, \epsilon \rangle\}$$

Algorithm 2 BACKWARDREACHABILITYDFS(P)

1: *Stack* := P;
2: *BRS* := P;
3: **while** *Stack* $\neq \emptyset$ **do**
4:     s := POP(*Stack*);
5:     **for** $s' \in \text{PRE}(s)$ **do**
6:         **if** $s' \notin BRS$ **then**
7:             BRS := $BRS \cup \{s'\}$;
8:             PUSH(*Stack*, s');
9:         **end if**
10:     **end for**
11: **end while**
12: **return** *BRS*;

Since assertion verification can be reduced to reachability checks as we discussed earlier, we can use the reachability algorithm above for verifying assertions. This approach is called *explicit state verification* since states of the transition system are visited individually. One of the problems with this approach is, for large state spaces, exploring state space one state at a time is computationally very expensive. In fact, as we observed, for string manipulating programs, the state space is infinite since we allow strings of arbitrary length. For infinite state systems explicit state verification cannot be used to prove absence of errors, but it can be used to prove existence of errors (since a trace that is discovered by explicit state enumeration that leads to an error state proves the existence of an error).

Explicit state verification can be used to guarantee absence of errors in finite state systems. For example, if we bound the variable domains in string programs we can use explicit state verification to explore the whole state space. However, there is another problem. Although depth first search algorithm explores the state space in linear time with respect to the size of the transition system (where the size of the transition system $T = (S, I, R)$ is $|S| + |T|$), the size of the transition system is exponential in the number of variables in the input program. The exponential growth of the state space of programs is called the *state space explosion problem*, and it limits the scalability of explicit state verification techniques for finite state systems.

3.3 Symbolic Exploration

As an alternative to explicit state enumeration we can consider exploring the state space using sets of states. Rather than exploring one state at a time, we will consider exploring sets of states. In order to do this, we need to first generalize the definition of pre and post-condition functions to sets of states as follows: PRE : $2^S \rightarrow 2^S$, POST : $2^S \rightarrow 2^S$, where

$$
\begin{aligned}
\text{POST}(P, l) &= \{s \mid \exists s' \in P \ : \ (s', s) \in r_l\} \\
\text{POST}(P) &= \{s \mid \exists s' \in P \ : \ (s', s) \in R\} \\
\text{PRE}(P, l) &= \{s \mid \exists s' \in P \ : \ (s, s') \in r_l\} \\
\text{PRE}(P) &= \{s \mid \exists s' \in P \ : \ (s, s') \in R\}
\end{aligned}
$$

We refer to POST and PRE as post-condition (or post-image) or pre-condition (or pre-image) functions.

For example, for the string program example shown in Fig. 3.1, we have the following:

$$
\begin{aligned}
\text{POST}(\{\langle 1, \epsilon, \epsilon, \epsilon \rangle\}) &= \{\langle 2, ab, \epsilon, \epsilon \rangle\} \\
\text{POST}(\{\langle 2, ab, \epsilon, \epsilon \rangle\}) &= \{\langle 3, ab, cd, \epsilon \rangle\} \\
\text{POST}(\{\langle 1, \epsilon, \epsilon, \epsilon \rangle, \langle 2, ab, \epsilon, \epsilon \rangle\}) &= \{\langle 2, ab, \epsilon, \epsilon \rangle, \langle 3, ab, cd, \epsilon \rangle\}
\end{aligned}
$$

The set of states can be infinite and using the set notation we can define the post-condition of an infinite set of states. For example, for the string program example shown in Fig. 3.2, we have the following:

$$\text{POST}(\{s \mid s = \langle 3, x, \epsilon, 0 \rangle\}) = \{s' \mid s' = \langle 4, x, \text{charat}(x, 0), 0 \rangle\}$$
$$\text{POST}(\{s \mid s = \langle 3, x, y, i \rangle\}) = \{s' \mid s' = \langle 4, x, y.\text{charat}(x, i), i \rangle\}$$
$$\text{POST}(\{s \mid s = \langle 4, x, y, \text{length}(y) - 1 \rangle\}) = \{s' \mid s' = \langle 4, x, y, \text{length}(y) \rangle\}$$

3.3.1 Symbolic Reachability

In order to explain forward and backward reachability computations on sets of states, we first define the lattice formed by the sets of states of the transition system.

Symbolic reachability algorithms deal with sets of states rather than individual states. By processing multiple states at the same time, symbolic techniques can converge to an answer with fewer iterations. For example, for the forward reachability analysis, if we want to compute the set of states reachable from the set of initial states, we can first start with the initial states I. Then we can add all the states reachable from initial states and continue adding new states until there is nothing new to add. This is exactly what the depth-first-traversal algorithm shown in Algorithm 1 does, but it does the traversal one state at a time. Symbolic reachability algorithms compute post-condition of a set of states in each iteration instead of computing post-condition of one state at a time. With an appropriate symbolic representation, symbolic algorithms can compute the post-condition of an infinite set of states in a single iteration.

The sets of states of a transition system form a partial order with respect to the set inclusion (i.e., $\subseteq$). The progress in reachability computations can be expressed with respect to this partial order. For the forward reachability computation, we start with I, and if we are using a symbolic representation, in the next iteration we would compute $I \cup \text{POST}(I)$. Note that $I \subseteq I \cup \text{POST}(I)$. We started with reachable states I and we made some progress by computing a potentially larger set of states in the next iteration. The goal of a forward symbolic reachability algorithm for computing reachable states would be to compute a larger set of states (with respect to the partial order) in each iteration and hopefully converge on the set of reachable states RS after a number of iterations.

These concepts about forward and backward reachability computations can be formalized by defining a lattice formed by the sets of states of the transition system. Given a transition system $T = (S, I, R)$, the power set of S, 2^S forms a complete lattice $(2^S, S, \emptyset, \subseteq, \cup, \cap)$, with the top element $\top = S$, the bottom element $\bot = \emptyset$, intersection $\cap$ as the meet (greatest lower bound) operator, union $\cup$ as the join (least upper bound) operator, and the set containment $\subseteq$ as the partial order. Then, PRE and POST are functions that map elements of this lattice (sets of states) to the elements of this lattice (sets of states).

Let us consider the string program example shown in Fig. 3.1. Here are some of the set of states for this program:

$$
\begin{array}{ll}
I & = \{\langle 1, \epsilon, \epsilon, \epsilon \rangle\} \\
\text{Post}(I) & = \{\langle 2, ab, \epsilon, \epsilon \rangle\} \\
I \cup \text{Post}(I) & = \{\langle 1, \epsilon, \epsilon, \epsilon \rangle, \langle 2, ab, \epsilon, \epsilon \rangle\} \\
\text{Post}(I \cup \text{Post}(I)) & = \{\langle 2, ab, \epsilon, \epsilon \rangle, \langle 3, ab, cd, \epsilon \rangle\} \\
I \cup \text{Post}(I \cup \text{Post}(I)) & = \{\langle 1, \epsilon, \epsilon, \epsilon \rangle, \langle 2, ab, \epsilon, \epsilon \rangle, \langle 3, ab, cd, \epsilon \rangle\} \\
\text{Post}(I \cup \text{Post}(I \cup \text{Post}(I))) & = \{\langle 2, ab, \epsilon, \epsilon \rangle, \langle 3, ab, cd, \epsilon \rangle\} \\
I \cup \text{Post}(I \cup \text{Post}(I \cup \text{Post}(I))) & = \{\langle 1, \epsilon, \epsilon, \epsilon \rangle, \langle 2, ab, \epsilon, \epsilon \rangle, \langle 3, ab, cd, \epsilon \rangle\}
\end{array}
$$

and here is how these sets are related in terms of the partial order $\subseteq$:

$$I \subseteq I \cup \text{Post}(I) \subseteq I \cup \text{Post}(I \cup \text{Post}(I)) \subseteq I \cup \text{Post}(I \cup \text{Post}(I \cup \text{Post}(I)))$$

Moreover, we can observe the following:

$$RS = I \cup \text{Post}(I \cup \text{Post}(I))$$
$$RS = I \cup \text{Post}(RS)$$

We see that the set of reachable states RS is the limit of the sequence of states we have been computing using the Post function. RS is greater than or equal to any element in the sequence, and, once we reach RS, the sequence stops increasing with respect to the partial order. We can explain these phenomena using the concept of fixpoints.

3.3.2 Fixpoints

Given a function $\mathcal{F} : 2^S \to 2^S$, let $\mathcal{F} P$ denote the application of function $\mathcal{F}$ to set $P \subseteq S$.

Given a function $\mathcal{F}$, x is called a fixpoint of the function if

$$\mathcal{F}x = x$$

Interestingly, as we show below, reachability properties can be expressed as fixpoints [136].

We use the lambda calculus notation for functions. A function with argument x is written in lambda calculus as follows: $\lambda x . \mathcal{F} x$

Consider the following function:

$$\lambda x . I \cup \text{Post}(x)$$

The set of reachable states RS is a fixpoint of this function, i.e.,

$$RS = I \cup \text{Post}(RS)$$

We can see this as follows: First, $RS \supseteq I \cup \text{Post}(RS)$ since $I \subseteq RS$, and any state reachable from a reachable state should be reachable itself, i.e., $\text{Post}(RS) \subseteq RS$. Next, we need to show that $RS \subseteq I \cup \text{Post}(RS)$. According to the definition of RS, the only way a state s can be in RS is, either 1) $s \in I$, or 2) there exists a state in RS that can reach s, which implies that $RS \subseteq I \cup \text{Post}(RS)$.

Next, we are going to show that RS is in fact the least fixpoint of this function. I.e., RS is the smallest fixpoint of the function $\lambda x . I \cup \text{Post}(x)$ with respect to the partial order $\subseteq$.

Let $\mu x . \mathcal{F} x$ denote the least fixpoint of $\mathcal{F}$, i.e., the smallest x such that $\mathcal{F} x = x$. Then, we claim that:

$$RS = \mu x . I \cup \text{Post}(x)$$

Note that, since RS is a fixpoint of the function $\lambda x . I \cup \text{Post}(x)$, and since $\mu x . I \cup \text{Post}(x)$ is the least fixpoint of the function $\lambda x . I \cup \text{Post}(x)$ we conclude that $\mu x . I \cup \text{Post}(x) \subseteq RS$.

Next, we need to prove that $RS \subseteq \mu x . I \cup \text{Post}(x)$ to complete the proof. Suppose z is a fixpoint of $\lambda x . I \cup \text{Post}(x)$. Then we know that $z = I \cup \text{Post}(z)$, which means that $\text{Post}(z) \subseteq z$. So, all states that are reachable from z are in z. Since we also have $I \subseteq z$, any path that is reachable from I must also be in z, which means that $RS \subseteq z$.

Since we showed that RS is contained in any fixpoint of the function $\lambda x . I \cup \text{Post}(x)$, it should also be contained in its least fixpoint, since the least fixpoint itself is a fixpoint. So we conclude that $RS \subseteq \mu x . I \cup \text{Post}(x)$ which concludes the proof.

Now, we discuss how to compute the least fixpoint. We call a function $\mathcal{F}$ *monotonic*, if $p \subseteq q$ implies $\mathcal{F} p \subseteq \mathcal{F} q$. We have the following property from the lattice theory [102]:

Let $\mathcal{F} : 2^S \rightarrow 2^S$ be a monotonic function. Then $\mathcal{F}$ always has a least fixpoint, which is defined as

$$\mu x . \mathcal{F} x \equiv \bigcap \{ x \mid \mathcal{F} x \subseteq x \}$$

Since $\mu x . \mathcal{F} x$ is the least fixpoint of the function $\mathcal{F}$, it is the intersection (greatest lower bound) of all the fixpoints of $\mathcal{F}$. In fact, it is the intersection of all the fixpoints of $\mathcal{F}$, i.e., it is the intersection of all the sets x where $\mathcal{F} x \subseteq x$. This property is valid even when S (hence the lattice) is infinite.

Given a function $\mathcal{F}$, $\mathcal{F}^i x$ is defined as:

$$\mathcal{F}^i x \text{ is defined as } \underbrace{\mathcal{F} (\mathcal{F} \dots (\mathcal{F} x))}_{i \text{ times}} .$$

We define $\mathcal{F}^0$ as the identity relation. Then, we have the following property [102]:
Given a monotonic function $\mathcal{F} : 2^S \to 2^S$, for all n,

$$\mu x \,.\, \mathcal{F} \, x \supseteq \bigcup_{i=0}^{n} \mathcal{F}^i \, \emptyset$$

This property holds even when the lattice is infinite.

Assume that we generate a sequence of approximations to the least fixpoint $\mu x \,.\, \mathcal{F} \, x$ of a monotonic function $\mathcal{F}$ by generating the following sequence:

$$\emptyset, \; \mathcal{F} \, \emptyset, \; \mathcal{F}^2 \, \emptyset, \; \ldots, \; \mathcal{F}^i \, \emptyset, \; \ldots$$

This sequence is monotonically increasing since $\emptyset$ corresponds to the bottom element of the lattice, and the function $\mathcal{F}$ is monotonic. If this sequence converges to a fixpoint, i.e., if we find an i where $\mathcal{F}^i \, \emptyset \equiv \mathcal{F}^{i+1} \, \emptyset$, then from the property above, we know that it is the least fixpoint, i.e., it is equal to $\mu x \,.\, \mathcal{F} \, x$.

Similarly, a monotonically decreasing sequence of approximations could be generated to compute the greatest fixpoint of a function [136]. In this monograph we are focusing on least fixpoints. Because of the duality between the least and the greatest fixpoints, the techniques described here can also be applied to computation of greatest fixpoint.

As an example for computing least fixpoints, consider the computation of the least fixpoint for reachable states: $RS = \mu x \,.\, I \cup \textsc{Post}(x)$. We can compute this least fixpoint by generating the following sequence:

$$\underbrace{\underbrace{\underbrace{\emptyset \vee I}_{\mathcal{F}\,\emptyset} \vee \textsc{Post}(I)}_{\mathcal{F}^2\,\emptyset} \vee \textsc{Post}\,(\textsc{Post}(I)) \vee \textsc{Post}\,(\textsc{Post}\,(\textsc{Post}(I))) \vee \ldots}_{\mathcal{F}^3\,\emptyset}$$

When this sequence converges to a fixpoint, the result will be equal to RS. This is exactly the sequence we computed for the string program example shown in Fig. 3.1 above.

In Algorithm 3 we give the fixpoint computation algorithm for the reachable states based on this iterative approach. Note that this fixpoint computation is closely related to the state space exploration algorithm given in Algorithm 1. Both algorithms first add the initial states to the reachable states and then keep adding states that are reachable from the initial states to the reachable states. They both stop when there is no more state left to add (i.e., when exploration reaches a fixpoint). The fixpoint computation algorithm traverses the state space in breadth-first order instead of the depth-first traversal order used in Algorithm 1. Also the fixpoint exploration algorithm processes a set of states at each iteration whereas the explicit state exploration algorithm processes a single state at each iteration.

Algorithm 3 REACHABILITYFIXPOINT

1: $RS := I$;
2: **repeat**
3:     $RS' := RS$;
4:     $RS := RS \cup \text{POST}(RS)$;
5: **until** $RS = RS'$
6: **return** RS;

Algorithm 4 BACKWARDREACHABILITYFIXPOINT(P)

1: $BRS := P$;
2: **repeat**
3:     $BRS' := BRS$;
4:     $BRS := BRS \cup \text{PRE}(BRS)$;
5: **until** $BRS = BRS'$
6: **return** BRS;

We can also compute backward reachability similarly using the PRE function as shown in Algorithm 4.

In order to implement the fixpoint computation algorithms we need a way to represent the sets of states. In general this representation should support tests for equivalence, emptiness, and membership, and meet (intersection) and join (union) operations. If the state space is finite, explicit state enumeration would be one such representation. Note that as the state spaces of the programs grow, explicit state enumeration will become more expensive since the size of this representation is linearly related to the number of states in the set it represents. Unfortunately, as we discussed above, the state spaces of programs increase exponentially with the number of variables. This state space explosion problem makes a naive implementation of the explicit state enumeration infeasible. Moreover, as we have seen for string programs, if we want to represent all possible string values during reachability analysis, then the number of states becomes infinite and an explicit state representation becomes impossible.

The symbolic reachability analysis techniques use a *symbolic representation* for encoding sets of states. Symbolic representations are mathematical objects (such as formulas in some logic) with semantics corresponding to sets of states. We can use such representations in encoding the sets of program states. Using a symbolic representation we can implement the iterative fixpoint computation algorithm and compute the reachable states. As we discuss in the next chapter, in this monograph we mainly focus on use of automata as a symbolic representation for sets of states of string programs.

3.4 Summary

In this chapter we provided a basic survey of reachability analysis for verification of string manipulating programs starting with explicit state enumeration. We discussed both forward and backward reachability analysis using depth-first search where states of a given string manipulating program are traversed one state at a time. Next, we discussed symbolic reachability analysis, where the basic idea is to perform state exploration using sets of states rather than traversing states one by one. We discussed that reachability analysis corresponds to fixpoint computations, and, in order to develop a symbolic analysis framework for string manipulating programs, we need to first develop a symbolic representation that can represent sets of strings. We discuss a symbolic representation for sets of strings in the next chapter.

Chapter 4
Automata Based String Analysis

In this chapter, we discuss using automata as a symbolic representation for string analysis. To compute forward and backward reachability for string manipulating programs, we can use automata-based symbolic string analysis where automata are used as a symbolic representation to represents sets of states of the program. We can iteratively compute an approximation of the least fixpoint that corresponds to the reachable values of the string expressions. Assume that we use one Deterministic Finite Automaton (DFA) per string variable, per program point. i.e., each DFA represents the set of values that a string variable can take at a particular program point. In each iteration, given the current state DFA for a variable, we can compute the pre- and post-state DFA. In order to implement this approach we have to develop automata based algorithms for computing pre- and post-state computation for common string operations such as concatenation, and replacement as we discuss later in this chapter.

4.1 A Lattice for Sets of Strings

Given an automaton A, let $\mathcal{L}(A)$ denote the set of values accepted by A. We focus on minimized, deterministic finite automata (DFA) and define the following partial order on automata: $A_1 \sqsubseteq A_2$ if and only if $\mathcal{L}(A_1) \subseteq \mathcal{L}(A_2)$. Given two automata A_1 and A_2, we define $A_1 \sqcup A_2$ as an automaton such that $\mathcal{L}(A_1 \sqcup A_2) = \mathcal{L}(A_1) \cup \mathcal{L}(A_2)$, and we define $A_1 \sqcap A_2$ as an automaton such that $\mathcal{L}(A_1 \sqcap A_2) = \mathcal{L}(A_1) \cap \mathcal{L}(A_2)$.

Given a regular expression r, we use $A(r)$ to denote an automaton that accepts the set of strings that match the regular expression r. Similarly, given a set of strings $P \subseteq \Sigma^*$, we use $A(P)$ to denote an automaton where $\mathcal{L}(A(P)) = P$. For example, the automaton $A(\Sigma^*)$ is an automaton that accepts every string, i.e., $\mathcal{L}(A(\Sigma^*)) = \Sigma^*$, and $A(\emptyset)$ is an automaton which does not accept any strings, i.e., $\mathcal{L}(A(\emptyset)) = \emptyset$.

© Springer International Publishing AG 2017

T. Bultan et al., *String Analysis for Software Verification and Security*,
https://doi.org/10.1007/978-3-319-68670-7_4

All the operators we defined above can be implemented using standard construction algorithms from automata theory.

Using the partial order $\sqsubseteq$ we can define an automata lattice with the join (least upper bound) operator $\sqcup$, the meet (greatest lower bound) operator $\sqcap$, the bottom element $\bot = A(\emptyset)$ and top element $\top = A(\Sigma^*)$. Note that the automata lattice is an infinite lattice and has infinite chains such as $\Sigma = \{a\}$, $A(\emptyset) \sqsubseteq A(\{\epsilon\}) \sqsubseteq A(\{\epsilon, a\}) \sqsubseteq A(\{\epsilon, a, aa\}) \sqsubseteq A(\{\epsilon, a, aa, aaa\}) \cdots \sqsubseteq A(a^*)$. Hence, fixpoint computations during symbolic reachability analysis are not guaranteed to converge when we use automata as a symbolic representation.

One approach for achieving convergence during fixpoint computations on lattices with infinite chains, is to use a widening operator ∇ to compute an over-approximation of the least fixpoint. The widening operator ∇ over-approximates the join operator, and it guarantees convergence. i.e., given two automata A_1 and A_2, $A_1 \sqcup A_2 \sqsubseteq A_1 \nabla A_2$, and the iterative least fixpoint computations are guaranteed to converge if we apply the widening operator at each iteration. After discussing symbolic analysis with automata as a symbolic representation, in the following chapters we discuss use of abstraction and approximation techniques, such as the widening operator, in order to achieve convergence.

4.2   Symbolic Reachability Analysis with Automata

In the following discussion we represent each program with a set of labels L, where each statement is marked with a unique label, and a set of flow edges $F \subseteq L \times L$ that denote the control flow among the statements of the program. Let V denote the set of string variables in the program, and let $l_1 \in L$ denote the first statement of the program.

We use an automata matrix of size $|L| \times |V|$ to represent the states of the program. Given a program location l and a variable v, and the automata matrix $\vec{A}$, the set of strings accepted by automaton $\vec{A}[l, v]$, i.e., $\mathcal{L}(\vec{A}[l, v])$ denotes the set of values that variable v can take at program location l. Given a $|L| \times |V|$ automata matrix $\vec{A}$, $\vec{A}[l]$ denotes the automata vector of size $|V|$ that corresponds to the l'th row of the automata matrix, which represents the set of values that variables can take at program location l.

We can lift the partial order we defined above for automata to automata vectors as follows: $\vec{A_1} \sqsubseteq \vec{A_2}$ if and only if $\forall v : \mathcal{L}(\vec{A_1}[v]) \subseteq \mathcal{L}(\vec{A_2}[v])$. Again we can also lift join, meet and widening operators to vectors of automata as follows:

$$\forall v : (\vec{A_1} \sqcup \vec{A_2})[v] = A_1[v] \sqcup A_2[v],$$
$$\forall v : (\vec{A_1} \sqcap \vec{A_2})[v] = A_1[v] \sqcap A_2[v],$$
$$\forall v : (\vec{A_1} \nabla \vec{A_2})[v] = A_1[v] \nabla A_2[v].$$

Algorithm 1 FORWARDANALYSIS(L, F, V)

1: $I := \{l \mid \forall l'.(l', l) \notin F\}$;
2: **for** $l \in L \setminus I, v \in V$ **do**
3:     $\vec{A}[l, v] = A(\emptyset)$;
4: **end for**
5: **for** $l \in I, v \in V$ **do**
6:     $\vec{A}[l, v] = A_{init}(v)$;
7: **end for**
8: queue $WQ := NULL$;
9: WQ.enqueue(l_1);
10: **while** $WQ \neq NULL$ **do**
11:     $l := WQ$.dequeue();
12:     **for** $(l, l') \in F$ **do**
13:         **if** POST($\vec{A}[l], (l, l')) \not\sqsubseteq \mathcal{L}(\vec{A}[l'])$ **then**
14:             $\vec{A}(l') = \vec{A}(l') \nabla (\vec{A}(l') \sqcup \text{POST}(\vec{A}(l), l))$;
15:             WQ.enqueue(l');
16:         **end if**
17:     **end for**
18: **end while**

Hence, this defines a lattice for automata vectors with the bottom element

$$\bot = \vec{A}(\emptyset), \quad \text{where } \forall v : \vec{A}(\emptyset)[v] = A(\emptyset)$$

and the top element

$$\top = \vec{A}(\Sigma^*), \quad \text{where } \forall lv : \vec{A}(\Sigma^*)[v] = A(\Sigma^*)$$

Algorithm 1 computes the least fixpoint that over-approximates the possible values that string variables can take at any given program point [127, 129]. It is a symbolic reachability computation that uses automata as a symbolic representation.

The algorithm first defines the initial set I that corresponds to the set of labels without incoming edges in F. These labels correspond to values that can be initialized from the start such as user inputs (which can be assumed to take any possible value) and constant strings. For labels in I, we assume that the algorithm initializes the corresponding automata vectors according to the type of the statement and the variables involved in the statement. For labels that are not in I, the algorithm initializes the automata in each corresponding automata vector to $A(\emptyset)$ which is the automaton that accepts no strings. i.e., initially we are assuming that no string value is reachable. Then, the algorithm iteratively propagates the reachable string values starting with the initial statements.

This algorithm is a worklist algorithm that stores statements to be processed in a worklist, which is implemented as a queue. First, we put the first statement l_1 to the worklist. At each iteration one statement is removed from the worklist, its post-condition is computed, and all successor statements are updated accordingly. When the set of strings reachable at a successor statement changes due to the

---

Algorithm 2 BACKWARDANALYSIS(L, F, V)

1: $T := \{l \mid \forall l'.(l, l') \notin F\}$;
2: **for** $l \in L \setminus T, v \in V$ **do**
3:     $\vec{A}[l, v] = A(\emptyset)$;
4: **end for**
5: **for** $l \in T, v \in V$ **do**
6:     $\vec{A}[l, v] = A_{init}(v)$;
7: **end for**
8: queue $WQ := NULL$;
9: WQ.enqueue(l_t);
10: **while do** $WQ \neq NULL$
11:     $l := WQ$.dequeue();
12:     **for** $(l, l') \in F$ **do**
13:         **if** PRE($\vec{A}[l], (l, l')) \not\sqsubseteq \mathcal{L}(\vec{A}[l'])$ **then**
14:             $\vec{A}(l') = \vec{A}(l') \nabla (\vec{A}(l') \sqcup$ PRE($\vec{A}(l), l$));
15:             WQ.enqueue(l');
16:         **end if**
17:     **end for**
18: **end while**

---

post-condition computation, then that successor statement is added to the worklist. This continues until reachable string values stop changing, and, hence, the worklist becomes empty. Note that, in order to guarantee convergence the algorithm uses the widening operator ∇.

Similar to the forward analysis, one can define the backward analysis to compute all potential input values that lead to the results at any given program point.

Algorithm 2 computes the least fixpoint that over-approximates the possible values that string variables can take that would lead to the results specified in the target label l_t [126, 128]. The terminal set T is defined as the set of terminal labels in F. These labels correspond to the automata matrix elements that can be initialized as arguments for pre-condition computations. For example, these automata can be computed via forward analysis to accept all reachable strings for the given program point [126, 128].

This backward analysis algorithm is also a worklist algorithm that stores statements to be processed in a worklist, which is implemented as a queue. First, we put the target statement (l_t) to the worklist. At each iteration one statement is removed from the worklist, its pre-condition is computed, and predecessor statements are updated accordingly. When the set of strings reachable at a predecessor statement changes due to the pre-condition computation, then that predecessor statement is added to the worklist. This continues until reachable string values stop changing, and, hence, the worklist becomes empty. The algorithm also adopts the widening operator ∇ to guarantee convergence.

In order to implement the Algorithms 1 and 2 we need to implement (1) a data structure for the automata that supports the join and widening operations, (2) can be used to check the partial order among the automata, and (3) supports post-condition and pre-condition operations. We discuss these below.

4.3 Symbolic Automata Representation

In order to reduce the memory usage during reachability analysis, we can use a symbolic DFA representation. The basic idea is to represent the transition relation of the automata symbolically using Multi Terminal Binary Decision Diagrams (MTBDDs) [41]. In order to do this, we first have to use a binary encoding of the set of characters that can appear in a string, i.e., the alphabet Σ.

Given $\mathcal{B} = \{0, 1\}$, a symbolic DFA A is a tuple $\langle Q, q_0, \Sigma_\mathcal{B}, \delta, F \rangle$ where:

- Q is a finite set of states.
- q_0 is the initial state.
- Instead of using a regular alphabet Σ, where each character c is a single ASCII printable symbol such as a and b, we use a symbolic binary alphabet $\Sigma_\mathcal{B} \subseteq \mathcal{B}^k$ where $k = log_2(\lceil |\Sigma| \rceil)$ and each alphabet symbol α is a k-bit string $\alpha \in \mathcal{B}^k$. In our discussion in this section, we use *character* to refer to a non-symbolic alphabet character $c \in \Sigma$ and we use *alphabet symbol* to refer to a symbolic alphabet symbol $\alpha \in \Sigma_\mathcal{B}$. Each character $c \in \Sigma$ is mapped to one and only one corresponding alphabet symbol $\alpha_c \in \Sigma_\mathcal{B}$ and vice versa.
- $F \subseteq Q$ is finite set of accepting states.
- $\delta : Q \times \Sigma_\mathcal{B} \to Q$ is the transition relation.

Following our definition of Σ and $\Sigma_\mathcal{B}$, we define a non-symbolic string w of length n as a sequence of characters $\langle c_0, c_1, \ldots, c_{n-1} \rangle$ where each character $c_i \in \Sigma$ and its corresponding symbolic string $w_\mathcal{B}$ as a sequence of alphabet symbols $\langle \alpha_0, \alpha_1, \ldots, \alpha_{n-1} \rangle$ where each alphabet symbol $\alpha_i \in \Sigma_\mathcal{B}$.

Let us define the relation $\delta^* : Q \times \Sigma_\mathcal{B}^* \to Q$ for the symbolic DFA A as follows: $\delta^*(q_i, \epsilon) = q_i$, and given a string $w_\mathcal{B} = \langle \alpha_0, \alpha_1, \ldots, \alpha_{n-1} \rangle$ where each character $\alpha_i \in \Sigma_\mathcal{B}$, $\delta^*(q_i, w_\mathcal{B}) = q_j$ if there exists a sequence $\langle q_i, q_{i+1}, q_{i+2}, \ldots, q_{i+n} \rangle \in Q^{n+1}$ such that: (1) $q_{i+n} = q_j$ and (2) $\forall 0 \le l < n : \delta(q_{i+l}, \alpha_l) = q_{i+l+1}$.

A state q of A is a *sink* state if $\forall \alpha \in \Sigma_\mathcal{B}, \delta(q, \alpha) = q$ and $q \notin F$. In the following discussion, we assume that for all unspecified pairs (q, α), $\delta(q, \alpha)$ goes to a *sink* state. When visualizing a DFA we omit the *sink* state and the transitions that lead to a sink state. We say that a string $w_\mathcal{B}$ is accepted by A if $\delta^*(q_0, w_\mathcal{B}) \in F$. The language of A or $\mathcal{L}(A) \subseteq \Sigma_\mathcal{B}^*$ is the set of strings $w_\mathcal{B}$ that are accepted by A. For two states q_i and q_j in A, a transition between q_i and q_j on an alphabet symbol α is a tuple (q_i, α, q_j) where $\delta(q_i, \alpha) = q_j$ and we write it like this $(q_i \xrightarrow{\alpha} q_j)$. For two states q_i and q_j in A, we say there is a path $q_i, q_{i+1}, \ldots, q_j$ of length n between q_i and q_j if there is a string $w_\mathcal{B} \in \Sigma_\mathcal{B}^*$ of length n such that $\delta^*(q_i, w_\mathcal{B}) = q_j$ and we write it like this $(q_i \xrightarrow{w_\mathcal{B}}{}^* q_j)$.

Fig. 4.1 Symbolic representation of a DFA using MTBDD

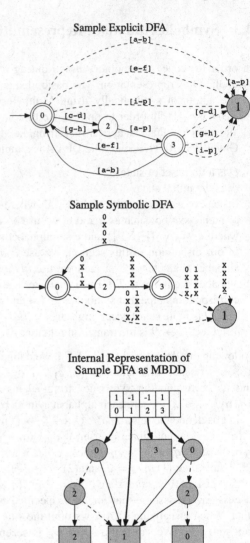

Sample Explicit DFA

Sample Symbolic DFA

Internal Representation of Sample DFA as MBDD

Example

Figure 4.1 shows an example symbolic DFA. At the top is the explicit DFA using an explicit representation that uses character ranges. In the middle is the symbolic DFA that uses binary alphabet symbols. In the bottom is the actual internal representation using a Multi-terminal Binary Decision Diagram (MTBDD). The alphabet Σ and the corresponding symbolic alphabet Σ_B is shown in Fig. 4.2. We have 16 characters in Σ which means that we need $log_2(\lceil 16 \rceil) = 4$ bits for the symbolic alphabet (i.e.,

```
a = 0000      b = 0001      0X0X = a, b, e, f = [a-b], [e-f]
c = 0010      d = 0011      0X1X = c, d, g, h = [c-d], [g-h]
e = 0100      f = 0101      1XXX = i, j, k, l, m, n, o, p = [i-p]
g = 0110      h = 0111      XXXX = a, b, c, d, e, f, g, h,
i = 1000      j = 1001             i, j, k, l, m, n, o, p = [a-p]
k = 1010      l = 1011
m = 1100      n = 1101
o = 1110      p = 1111
```

Fig. 4.2 Σ and corresponding Σ_B for the sample symbolic DFA along with symbolic transition labels and their corresponding explicit transitions

$\Sigma_B \subseteq B^4$). Figure 4.2 shows each character $c \in \Sigma$ and its corresponding alphabet symbol $\alpha_c \in \Sigma_B$. For example, $\alpha_a = 0000$ and $\alpha_l = 1011$.

The sample DFA in Fig. 4.1 has four states $Q = \{S_0, S_1, S_2, S_3\}$. Our convention is to label each state with a number n and refer to it in the text with S_n. The shaded state S_1 is the sink state and dashed edges represent transitions that go to sink state. From now on we always omit sink states to simplify the figures (shaded states and dashed edges will be used for other purposes). In the sample symbolic DFA in the middle, we use symbolic labels such as $\begin{smallmatrix}0\\X\\1\\X\end{smallmatrix}$ where the X symbol indicates that the value could be 0 or 1. For example, from state $S_0 \rightarrow S_2$, an edge labeled $\begin{smallmatrix}0\\X\\1\\X\end{smallmatrix}$ means that there are four transitions between these two states on alphabet symbols $0010, 0011, 0110$ and 0111 (i.e., four transitions on characters c, d, g, h). Figure 4.2 shows each of the symbolic labels used in sample symbolic DFA and their meanings.

4.3.1 MTBDD Representation

Figure 4.1 shows at the bottom the Multi-terminal Binary Decision Diagram [41] that is used as the actual internal representation of the sample symbolic DFA. The second row in table at the top $\boxed{0}\boxed{1}\boxed{2}\boxed{3}$ represents DFA states while the first row in that table $\boxed{1}\boxed{-1}\boxed{-1}\boxed{1}$ represents state types which are either accepting state 1 or rejecting state -1. The shaded nodes represent *BDD nodes*. Each circle-shaped node has a number n that represents its level i.e., which BDD variable n (in other words, which bit n in an alphabet symbol $\alpha \in \Sigma_B$) it corresponds to. Each rectangle-shaped node has a number n that represents the destination state S_n that the node corresponds to. Dashed line represents a BDD variable (bit) value of 0 while a regular line represents a BDD variable (bit) value of 1.

The symbolic transition relation works as follows: Suppose that we are in state S_0 and given alphabet symbol $\alpha_c = 0010$. Let us see how we go from state S_0 to state S_2 on character c. We start from table cell $\boxed{0}$ then go to a BDD node at level 0. Then, looking at value of bit-0 of α_c which is 0, we go to a BDD node at level 2.

Notice that since we are in a BDD node at level. 2, we will look at second BDD variable (i.e., bit-2) skipping value of bit-1 as it does not affect which destination state we will go to. Then we look at value of bit-2 which is 1 which means that we go to destination state S_2 (skipping value of bit-3).

Throughout the remaining of this text we use a representation of a DFA in a figure that is a mixture of the top two representations in Fig. 4.1. On one hand, we will have one edge only between each two states in the DFA (as we did in the middle sample symbolic DFA). On the other hand, instead of labeling the edge with a symbolic label like $\begin{smallmatrix} 1 \\ X \\ X \\ X \end{smallmatrix}$ we will label it with either character ranges such as [a-c] or a character set such as $\{a, b\}$ and $\Sigma - \{d, h, k\}$. In addition, we will always omit the sink state and all transitions that go to it.

4.3.2   Modeling Non-Determinism Using Symbolic DFA

The MTBDD-based symbolic automata representation does not support non-determinism directly. For a given alphabet symbol and an automata state, the MTBDD data structure will only point to one target state. Therefore, we need to implement the pre and post condition computations for string operations without using the standard constructions based on the ϵ-transitions since the MTBDD-based automata representation does not allow ϵ-transitions. We model non-determinism in the MTBDD-based automata representation by extending the alphabet with extra bits and then projecting them away using the on-the-fly subset construction algorithm. Projection is applied one bit at a time, and after projecting each bit away, the size of the resulting automaton is reduced using MTBDD-based automata minimization.

In other words, the project and determinize operation, denoted as PROJECT(A, i), where $1 \leq i \leq k$, converts a DFA A recognizing a language L over the alphabet $\Sigma_B \subseteq \mathcal{B}^k$, to a DFA A' recognizing a language L' over the alphabet $\Sigma_B \subseteq \mathcal{B}^{k-1}$, where L' is the language that results from applying the tuple projection on the i-th bit to each symbol of the alphabet. The process consists of removing the i-th track of the MTBDD and determinizing the resulting MTBDD via on-the-fly subset construction. If we have a DFA A recognizing a language L over the alphabet $\Sigma_B \subseteq \mathcal{B}^k$, and we want to add to A n non-deterministic transitions out from a state S on some character c, we need to extend Σ_B with $l = \lceil log_2(n) \rceil$ extra bits to get $\Sigma_B' \subseteq \mathcal{B}^{k+l}$. Then we determinize by projecting the extra l bits one bit at a time.

Figure 4.3 shows on the left part of a Non-deterministic FA with three non-deterministic transitions on character a from state S_0 to states S_1, S_2 and S_3. On the right it shows the corresponding symbolic DFA where $\Sigma_B \subseteq \mathcal{B}^2$ and $\alpha_a = 00$. To simulate non-determinism, we need to extend the alphabet Σ_B by adding two extra bits to represent three different symbols for character a namely a_0, a_1 and a_2 where $\alpha_{a_0} = 0000$, $\alpha_{a_1} = 0001$ and $\alpha_{a_2} = 0010$. At the end, we determinize the DFA by projecting bit-3 then bit-2.

Fig. 4.3 A symbolic DFA on the right with $\Sigma_\mathcal{B} \subseteq \mathcal{B}^2$ simulating non-determinism in the NFA on the left using two extra bits

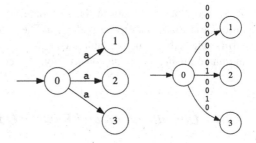

4.3.3   Symbolic vs. Explicit DFA

MTBDD-based symbolic DFA representation is more efficient in terms of memory than explicit DFA which means that, using symbolic DFA, our analysis consumes less memory. Although both explicit and symbolic representation of sample DFA shown in Fig. 4.1 seem to use the same number of transitions, bear in mind that a BDD transition is labeled with a single bit while an explicit DFA transition is labeled with 2 characters using 4 bits to represent each one ($4 = log_2(|\Sigma|)$). In general, the difference between explicit and symbolic DFA in memory consumption becomes more obvious as size of alphabet $|\Sigma|$ grows such as the case with $|\Sigma_{ASCII}| = 256$ and $|\Sigma_{UNICODE}| = 65{,}536$. Furthermore, as we will see in the following chapters, in order to perform relational string analysis, we use multi-track automata which increases the size of the alphabet for the automata. For multi-track automata, use of MTBDD-based symbolic DFA representation is crucial for memory efficiency.

4.4   Post-Condition Computation

In order to implement forward symbolic analysis, we need to compute post-conditions of statements. In order to define post-condition on the symbolic representation, we overload the POST operation, so that it not only works on sets of states but also on automata that represent sets of states. Given an automata vector $\vec{A}$ where $\mathcal{L}(\vec{A}[l])$ denotes the set of values that variables can have just before the execution of statement l, $\text{POST}(\vec{A}, (l, l'))$ is the automata vector that represents the set of values that the variables can take after the execution of the statement l and before the execution of the successor statement l'.

In this section we focus on computing the post-condition of assignment statements of the form: $v := sexp$; where $sexp$ is a string expression. Computing post-condition of branch statements are discussed in Chaps. 5 and 7. In order to compute the post-condition of these statements we need to construct automata for different string operations. In the next section, we discuss how to do that with two most common string operations: concat and replace. Note that, for the assignment statements, $\text{POST}(\vec{A}, (l, l'))$ only has to update one automaton in the automata vector

$\vec{Z}$, all the other automata in the automata vector will remain the same. The only automaton that needs to be updated is the automaton $A[v]$ that accepts the set of string values that string variable v can take. We discuss how to construct the updated automaton using automata constructions for concat and replace operations.

4.4.1 Concatenation and Replace Operations

String operations *concatenation* and *replacement* are commonly used in modern software applications to manipulate strings.

Consider the string expression x.y where x and y are two string variables. If the value of the variable x is string w_1 and the value of the variable y is string w_2, then the value of the expression x.y would be the string w_1w_2 corresponding to the concatenation of the strings w_1 and w_2. In string analysis we are interested in computing all possible values that string expressions can take. Hence, we are interested in computing the set of possible values for the string expression x.y given a set of string values for x and a set of string values for y. Let us assume that the set of string values for x is denoted by the language L_1 and the set of string values for y is denoted by the language L_2. Then, the set of values for the string expression x.y corresponds to the concatenation of the two languages L_1 and L_2 which is defined as the string set $\{w_1w_2 \mid w_1 \in L_1, w_2 \in L_2\}$. Since we use DFA to represent sets of strings, we say a DFA A is the concatenation-DFA of A_1 and A_2 if and only if A accepts the concatenation of $\mathcal{L}(A_1)$ and $\mathcal{L}(A_2)$, i.e.,

$$\mathcal{L}(A) = \{w_1w_2 \mid w_1 \in \mathcal{L}(A_1), w_2 \in \mathcal{L}(A_2)\}.$$

We are also interested in computing the set of values that the string expression replace(s, p, t) can have given sets of possible string values for s (the source string), p (the match pattern), and t (the target string). We extend the definition of the replace operation to languages as we did with the concatenation operation above. We define the replacement operation on languages as follows. Given A_1, A_2, and A_3 that accept the original strings (s in replace(s, p, t)), the match strings (p in replace(s, p, t)), and the replacement strings (t in replace(s, p, t)), respectively, the replacement language of the DFA tuple (A_1, A_2, A_3) is defined as the set

$$\{w \mid k > 0, w_1x_1w_2 \ldots w_kx_kw_{k+1} \in \mathcal{L}(A_1), w = w_1c_1w_2 \ldots w_kc_kw_{k+1},$$
$$\forall 1 \leq i \leq k, x_i \in \mathcal{L}(A_2), c_i \in \mathcal{L}(A_3),$$
$$\forall 1 \leq i \leq k+1, w_i \notin \{w_1'x'w_2' \mid x' \in \mathcal{L}(A_2), w_1', w_2' \in \Sigma^*\}\}$$

We say a DFA A is the replaced-DFA of a DFA tuple (A_1, A_2, A_3) if and only if A accepts the replacement language of the DFA tuple (A_1, A_2, A_3). That is, A accepts the set of strings generated from a string s accepted by A_1 whose substrings that are accepted by A_2 are all replaced with any string r accepted by A_3.

The `replace(s, p, t)` operation we defined in Chap. 2 is a generic replace operation and the language based replace operation we defined above generalizes it to sets of strings. There are many variations of replace operation semantics in different programming languages. For example, in PHP programs, replacement operations such as `ereg_replace` can use different replacement semantics such as *longest match* or *first match*. Our replacement operation provides an over approximation of such more restricted replace semantics. For example, consider $\mathcal{L}(A_1) = \{baab\}$, $\mathcal{L}(A_2) = a^+$ (A_2 accepts the language $\{a, aa, aaa, \ldots\}$) and $\mathcal{L}(A_3) = \{c\}$. According to the longest match semantics, A only accepts bcb, in which the longest match aa is replaced by c. In the first match semantics, A only accepts $bccb$, in which two matches a and a are replaced with c. Both of these are included in the result obtained by our replacement operation. This over approximation works well and does not raise too many false alarms in practice [127]. Indeed, we have observed that most statements in string manipulating programs yield the same result with respect to the first and longest match semantics, e.g., `ereg_replace("<script *>","",$_GET["username"]);`, which are precisely modeled by the language-based replacement operation. On the other hand, there have been techniques proposed for precise modeling of replace operation with respect to different matching strategies using transducers [91]. The replacement technique discussed below can be extended to different matching strategies using a similar approach.

4.4.2   Post-Condition of Concatenation

Assume that we want to compute the post-condition of the following assignment statement `z := x . y`. We assume that we are given two automata A_x and A_y characterizing the set of string values that string variables `x` and `y` can take before the assignment statement is executed. Given A_x and A_y, we define an operator PostConcat(DFA A_1, DFA A_2) that returns an automaton that accepts the set of string values that the string expression `x . y` can take at that program point. Then, we can compute the post-condition of the assignment statement by updating the automaton for variable `z` as:

$$A_z = \text{PostConcat}(A_x, A_y)$$

where A_z is the automaton that accepts the set of string values that string variable `z` can take after the execution of the assignment statement.

PostConcat(DFA A_1, DFA A_2) returns a DFA that accepts the concatenation of strings accepted by A_1 and A_2. Given $A_1 = \langle Q_1, q_{10}, \Sigma, \delta_1, F_1 \rangle$ and $A_2 = \langle Q_2, q_{20}, \Sigma, \delta_2, F_2 \rangle$, the concatenation-DFA $A = \text{PostConcat}(\text{DFA } A_1, \text{DFA } A_2)$ can be constructed as follows. Without loss of generality, we assume that $Q_1 \cap Q_2$ is empty. We first construct an intermediate DFA $A' = \langle Q', q_{10}, \Sigma', \delta', F' \rangle$, where

- $Q' = Q_1 \cup Q_2$
- $\Sigma' = \{\alpha 0 \mid \alpha \in \Sigma\} \cup \{\alpha 1 \mid \alpha \in \Sigma\}$
- $\forall q, q' \in Q_1, \delta'(q, \alpha 0) = q'$, if $\delta_1(q, \alpha) = q'$
- $\forall q, q' \in Q_2, \delta'(q, \alpha 0) = q'$, if $\delta_2(q, \alpha) = q'$
- $\forall q \in Q_1, \delta'(q, \alpha 1) = q'$, if $q \in F_1$ and $\exists q' \in Q_2, \delta_2(q_{20}, \alpha) = q'$
- $F' = F_1 \cup F_2$, if $q_{20} \in F_2$; F_2, otherwise.

Then, $A = \text{PROJECT}(A', k + 1)$. Since both A_1 and A_2 are DFA, the subset construction happens only when there exists $q \in F_1$ such that $\exists \alpha, q', q'', \alpha \in \Sigma, q' \in Q_1, q'' \in Q_2, \delta_1(q, \alpha) = q', \delta_2(q_{20}, \alpha) = q''$.

4.4.3   Post-Condition of Replacement

In order to compute the post condition of an assignment statement of the form z := replace(s, p, t) we use a similar function: POSTREPLACE.

POSTREPLACE(DFA A_1, DFA A_2, DFA A_3) returns a DFA that accepts the replaced strings where A_1 accepts the original strings, A_2 defines the match, and A_3 defines the replacement. The replaced-DFA $M = \text{POSTREPLACE}(\text{DFA } A_1, \text{DFA } A_2, \text{DFA } A_3)$ accepts strings that are obtained by taking a string that A_1 accepts and replacing all substrings that match $\mathcal{L}(A_2)$ with a string in $\mathcal{L}(A_3)$. Table 4.1 shows several examples for $M = \text{POSTREPLACE}(A_1, A_2, A_3)$. M accepts an over approximation with respect to the leftmost (first) and longest match replace operations. In practice, many string functions can be modeled using the replace operation. For example, PHP replace operations such as preg_replace and ereg_replace that have regular expressions as their arguments can be implemented using automata constructions similar to the one we describe below. Also, other PHP functions such as htmlspecialchars, tolower, toupper, str_replace, and trim can be modeled using the replace operation.

To construct the replaced-DFA $A = \text{POSTREPLACE}(A_1, A_2, A_3)$, we insert marks into automata, identify matching sub-strings by intersection of automata, and then construct the final automaton by replacing these matching sub-strings with replacement. The details of automata constructions can be found in [129].

Here we use a running example to illustrate the idea (see Fig. 4.4). Consider $\mathcal{L}(A_1) = \{baab\}$, $\mathcal{L}(A_2) = a^+$ (A_2 accepts the language $\{a, aa, aaa, \ldots\}$) and

Table 4.1 Examples of string replace operation

$\mathcal{L}(A_1)$	$\mathcal{L}(A_2)$	$\mathcal{L}(A_3)$	$\mathcal{L}(M)$
{baaabaa}	{aa}	{c}	{bacbc, bcabc}
{baaabaa}	a^+	ϵ	{bb}
{baaabaa}	$a^+ b$	{c}	{baacaa, bacaa, bcaa}
{baaabaa}	a^+	{c}	{bcccbcc, bcccbc,
			bccbcc, bccbc, bcbcc, bcbc}
$ba^+ b$	a^+	{c}	$bc^+ b$

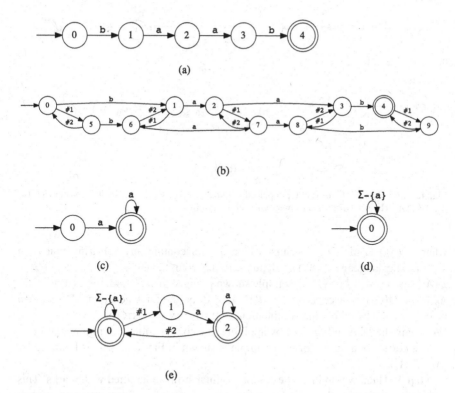

Fig. 4.4 Step 1 and Step 2: Add marks to sub strings and match. (**a**) A_1. (**b**) A_1': original strings with substrings marked. (**c**) A_2. (**d**) A_h. (**e**) A_2': arbitrary strings with match marked

$\mathcal{L}(A_3) = \{c\}$. Let $|A|$ denote the number of states of A. An upper bound for each intermediate automaton before projection and minimization is also described.

Step 1: We first insert marks to identify all sub strings of in $\mathcal{L}(A_1)$. We use the symbol $\sharp_1$ to denote the beginning of a substring and the symbol $\sharp_2$ to denote the end of a substring. Given A_1, we construct A_1', s.t.

$$\mathcal{L}(A_1') = \{w' \mid k > 0, w = w_1 x_1 w_2 \ldots w_k x_k w_{k+1} \in \mathcal{L}(A_1),$$
$$w' = w_1 \sharp_1 x_1 \sharp_2 w_2 \ldots w_k \sharp_1 x_k \sharp_2 w_{k+1}\}.$$

The example for constructing the automata A_1' with substrings of strings in $\mathcal{L}(A_1)$ identified with marks is shown in Fig. 4.4a and 4.4b. $|A_1'|$ is bounded by $2 \times |A_1|$.

Step 2: Next, we generate arbitrary strings with all substrings that are in $\mathcal{L}(A_2)$ identified with marks $\sharp_1$ and $\sharp_2$. Given the match automata A_2, we construct A_2', s.t.

$$\mathcal{L}(A_2') = \{w' \mid k > 0, w' = w_1 \sharp_1 x_1 \sharp_2 w_2 \ldots w_k \sharp_1 x_k \sharp_2 w_{k+1},$$
$$\forall 1 \leq i \leq k, x_i \in \mathcal{L}(A_2), \forall 1 \leq i \leq k+1, w_i \in \mathcal{L}(A_h)\},$$

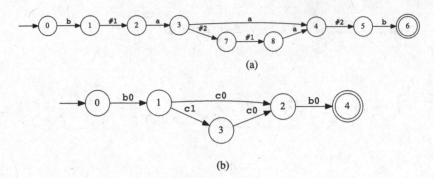

Fig. 4.5 Step 3 and 4: Construct the replaced automata by finding the path with replacement. (a) A': $A_1' \sqcap A_2'$. (b) A'': string replacement with auxiliary bits

where $\mathcal{L}(A_h)$ is the set of strings which do not contain any substring that is in $\mathcal{L}(A_2)$. The language $\mathcal{L}(A_h)$ is defined as the complement set of $\{w_1 x w_2 \mid x \in \mathcal{L}(A_2), w_1, w_2 \in \Sigma^*\}$. As the example shown in Fig. 4.4c and 4.4d, for $\mathcal{L}(A_2) = a^+$, A_h is the DFA that accepts $(\Sigma - \{a\})^*$. Let A^* be the DFA accepting Σ^*. A_h can be constructed by taking the complement of $\text{POSTCONCAT}(\text{POSTCONCAT}(A^*, A_2), A^*)$. We obtain the DFA in Fig. 4.4d by applying this construction with minimization.

The corresponding A_2' for our example is shown in Fig. 4.4e. $|A_2'|$ is bounded by $|A_h| + |A_2|$.

Step 3: Then, we find matches in the original strings identified with marks. This is done by computing the intersection language of A_1' and A_2' using automata product operation. The intersection automata $A' = A_1' \sqcap A_2'$ in our example is shown in Fig. 4.5a. $|A'|$ is bounded by $|A_1'| \times |A_2'|$.

Step 4: Finally, we replace all matches that are identified with marks with the replacement string. We first introduce a function $reach : Q \rightarrow 2^Q$, which maps a state to all its $\sharp$-reachable states in A. We say q' is $\sharp$-reachable from q if there exists w, $q' = \delta^*(q, \sharp_1 w \sharp_2)$. For instance, in Fig. 4.5, one can find that $reach(q_1) = \{q_5, q_7\}$ and $reach(q_7) = \{q_5\}$. Intuitively, one can think that each pair (q, q'), where $q' \in reach(q)$, identifies a word in $\mathcal{L}(A_2)$.

The idea to construct the final automaton from A' is to, for each q and $q' \in reach(q)$, insert paths between q and q' that recognize a string in $\mathcal{L}(A_3)$. To deal with nondeterminism with MTBDDs, we add extra bits to the alphabet as we did in the construction of concatenation. Extra bits are added to the alphabet to make transitions deterministic and later be projected away to yield the DFA with the original alphabet.

The final replaced-DFA A can then be constructed by iteratively projecting away the extra bits (over Σ) in Fig. 4.5b. The subset construction is only applied when needed.

The replaced-DFA of (A_1, A_2, A_3), where $\mathcal{L}(A_1) = \{baab\}$, $\mathcal{L}(A_2) = a^+$, and $\mathcal{L}(A_3) = \{c\}$, is A that accepts $\{bcb, bccb\}$.

The subset construction may induce an exponential blow-up of the number of states of the final DFA. In fact, it has been shown that a potential exponential blow-up of the number of states of the final DFA is inevitable in a replacement operation in some cases [129].

4.5 Pre-Condition Computation

In order to implement backward symbolic analysis, we need to compute pre-conditions of statements. Below, we explain how we compute the pre-conditions of the string operations which can then be used to compute pre-conditions of statements. We will focus on computing the pre-condition of concatenation and replacement operations as we did for post-condition computations. Specifically, we will discuss the functions PRECONCATPREFIX(DFA A, DFA A_2), PRECONCATSUF-FIX(DFA A, DFA A_1), and PREREPLACE(DFA A, DFA A_2, DFA A_3) which return a DFA:

- PRECONCATPREFIX(A, A_2) returns a DFA A_1 such that $A = $ POSTCONCAT(A_1, A_2).
- PRECONCATSUFFIX(A, A_1) returns a DFA A_2 such that $A = $ POSTCONCAT(A_1, A_2).
- PREREPLACE(A, A_2, A_3) returns a DFA A_1 such that $A = $ POSTREPLACE(A_1, A_2, A_3).

Note that, similar to post-condition computations, we may not be able to compute the precise pre-condition automaton. In such cases, we produce an over-approximation of the pre-condition automaton.

4.5.1 Pre-Condition of Concatenation

To compute the pre-condition of concatenation operation, we introduce concatenation transducers to specify the relation among its output and two input expressions. Transducers are multi-track automata we use for pre-condition computation (these are similar to the multi-track automata we use for relational string analysis we discuss in the next Chapter).

A concatenation transducer is a multi-track DFA over the alphabet that consists of 3 tracks (we discuss multi-track DFAs in more detail in the next chapter). The 3-track alphabet is defined as $\Sigma^3 = \Sigma \times (\Sigma \cup \{\lambda\}) \times (\Sigma \cup \{\lambda\})$, where $\lambda \notin \Sigma$ is a special symbol for padding. We use $w[i]$ $(1 \leq i \leq 3)$ to denote the ith track of $w \in \Sigma^3$. All tracks are aligned. $w[1] \in \Sigma^*$, $w[2] \in \Sigma^* \lambda^*$ is left justified, and $w[3] \in \lambda^* \Sigma^*$ is right justified. We use $w'[2], w'[3] \in \Sigma^*$ to denote the λ-free prefix of $w[2]$ and the λ-free suffix of $w[3]$. We say w is accepted by a concatenation transducer A if $w[1] = w'[2].w'[3]$. Since a concatenation transducer binds the values of different tracks character by character it is able to identify the prefix and suffix relations precisely.

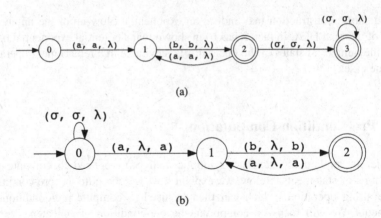

Fig. 4.6 Transducers for (**a**) $X = (ab)^+.Z$ and (**b**) $X = Y.(ab)^+$

Below we show two examples of concatenation transducers. Let α indicate any character in Σ. In Fig. 4.6a, the third track of A can be used to identify all suffixes of X that follow any string in $(ab)^+$. In Fig. 4.6b, the second track of A can be used to identify all prefixes of X that are followed by any string in $(ab)^+$.

In the following, we describe how to compute the pre-images of a concatenation node using concatenation transducers.

PreConcatPrefix(DFA A, DFA A_2) returns the prefix automata A_1 such that A = PostConcat(A_1,A_2). Consider the assignment statement z := x . y, in the pre-condition computation, we would like to compute the set of possible values for variable x given the set of possible values for z and for y. We can use the function PreConcatPrefix to compute the pre-condition automaton for variable x for the statement z := x . y.

Given $A = \langle Q_x, \Sigma, \delta_x, q_x0, F_x \rangle$ and $A_2 = \langle Q_z, \Sigma, \delta_z, q_{z0}, F_z \rangle$, PreConcatPrefix($A, A_2$) returns A_1 which is constructed using the following steps:

- Extend A to a 3-track DFA A', so that A' accepts $\{w \mid w[1] \in \mathcal{L}(A)\}$.
- Construct the concatenation transducer A_t that accepts $\{w \mid w[1] = w'[2].w'[3], w'[3] \in \mathcal{L}(A_2)\}$.
  $A_t = \langle Q, \Sigma^3, \delta, q_0, F \rangle$, where:

  - $Q = \{q_0\} \cup Q_2$,
  - $\forall a \in \Sigma, \delta(q_0, (a, a, \lambda)) = q_0$,
  - $\forall a \in \Sigma, \delta(q_0, (a, \lambda, a)) = q'$ if $\delta_z(q_{z0}, a) = q'$.
  - $\forall q, q' \in Q_2, \forall a \in \Sigma, \delta(q, (a, \lambda, a)) = q'$ if $\delta_z(q, a) = q'$.
  - $F = \{q_0\} \cup F_2$ if $q_{z0} \in F_2$. $F = F_2$, otherwise.

- Intersect A' with A_t. The result accepts $\{w \mid w[1] = w'[2].w'[3], w[1] \in \mathcal{L}(A), w'[3] \in \mathcal{L}(A_2)\}$. We then project away the first and the third tracks. Let the result be $A'_1 = \langle Q_1, \Sigma \cup \{\lambda\}, \delta, q'_{y0}, F'_1 \rangle$.

- Remove λ tails, if any, to construct A_1.
  $A_1 = \langle Q_1, \Sigma, \delta_1, q_{10}, F_1 \rangle$ as:

  - $\forall q, q' \in Q_1, \forall a \in \Sigma, \delta_1(q, a) = q'$ if $\delta_1'(q, a) = q'$.
  - $F_1 = F_1' \cup F_\lambda$, where $F_\lambda = \{q \mid \exists q' \neq sink, \delta_1'(q, \lambda) = q'\}$.

Similarly, PRECONCATSUFFIX(DFA A, DFA A_1) returns the suffix automata A_2 such that $A = $ POSTCONCAT(A_1, A_2). PRECONCATSUFFIX is used for computing the pre-condition automaton for variable x for the statement z := x . y.

Again, let $A = \langle Q, \Sigma, \delta, q_0, F \rangle$, and $A_1 = \langle Q_1, \Sigma, \delta_1, q_{10}, F_1 \rangle$. PRECONCATSUFFIX($A, A_1$) constructs the automaton A_2 using the following steps:

- Extend A to a 3-track DFA A', so that A' accepts $\{w \mid w[1] \in \mathcal{L}(A)\}$.
- Construct the concatenation transducer A_t that accepts $\{w \mid w[1] = w'[2].w'[3], w'[2] \in \mathcal{L}(A_1)\}$. $A_t = \langle Q_t, \Sigma^3, \delta_t, q_{10}, F_t \rangle$, where:

  - $Q_t = Q_1 \cup \{q_f\}$
  - $\forall q, q' \in Q_1, \forall a \in \Sigma, \delta(q, (a, a, \lambda)) = q'$ if $\delta_1(q, a) = q'$.
  - $\forall q \in F_1, \forall a \in \Sigma, \delta(q, (a, \lambda, a)) = q_f$.
  - $\forall a \in \Sigma, \delta(q_f, (a, \lambda, a)) = q_f$.
  - $F_t = \{q_f\} \cup F_1$.

- Intersect A' with A_t. The result $A' \sqcap A_t$ accepts $\{w \mid w[1] = w'[2].w'[3], w[1] \in \mathcal{L}(A_x), w'[2] \in \mathcal{L}(A_1)\}$. We then project away the first and the second tracks. Let the result be $A_2' = \langle Q_2', \Sigma \cup \{\lambda\}, \delta_2', q_{20}', F_2' \rangle$.
- Remove λ heads if any and construct $A_2 = \langle Q_2, \Sigma, \delta_2, q_{20}, F_2 \rangle$ as:

  - $Q_2 = q_0 \cup Q_2'$.
  - $\forall q \in Q_2', \forall a \in \Sigma, \delta_2(q, a) = q'$ if there exists $q' \in Q_2', \delta_2'(q, a) = q'$.
  - $\forall q \in Q_2', \forall a \in \Sigma, \delta_2(q_{20}, a) = q'$ if there exists $q', q'' \in Q_2', \delta_2'(q'', \lambda) = q$ and $\delta_2'(q, a) = q'$.
  - $F_2 = \{q_0\} \cup F_2'$, if $\exists q \in F_2'$ and there exists $q', q'' \in Q_2'$, so that $\delta_2'(q'', \lambda) = q$ and $\delta_2'(q, a) = q'$. $F_2 = F_2'$, otherwise.

4.5.2 Pre-Condition of Replacement

PREREPLACE(DFA A, DFA A_2, DFA A_3) returns the automaton A_1 such that $A = $ POSTREPLACE(A_1, A_2, A_3).

Recall that a `replace` statement has three inputs: the original strings, the match pattern, and the replacement pattern. Consider the replace statement in the form z := replace(s, p, t). Let us consider to compute all potential original strings (as the pre-image of the target strings) given regular sets characterizing possible values of the resulting strings, the match pattern, and the replacement pattern. Let $A_z = $ REPLACE(A_s, A_p, A_t), then our goal is to compute A_s, given A_z, A_p, and A_t.

An intuitive potential solution of PREREPLACE(A_z, A_p, A_t) is POSTREPLACE(A_z, A_t, A_p). However, since not all matches of A_t that appear in A_z are due to the replacement operation, this would be an incorrect solution. Consider a simple example. A_s, A_p and A_t accepts $\{aab\}$, $\{b\}$, and $\{a\}$, respectively. A_z = POSTREPLACE(A_s, A_p, A_t) accepts $\{aaa\}$. A'_s = POSTREPLACE(A_z, A_t, A_p) accepts $\{bbb\}$. Since $\{bbb\}$ does not include $\{aab\}$, this intuitive approach is not correct. To address this issue, a conservative solution is proposed in [129], where PREREPLACE(DFA A_z, DFA A_p, DFA A_t) is modeled as POSTREPLACE(A_z, A_t, $A_p \sqcup A_t$). The result is an over approximation of the pre-condition. Consider another simple example. A_s, A_p and A_t accept $\{a\}$, $\{a\}$, and $\{aa\}$, respectively. A_z = POSTREPLACE(A_s, A_p, A_t) accepts $\{aa\}$. A'_s = POSTREPLACE(A_z, A_t, $A_p \sqcup A_t$) accepts $\{a, aa\}$. In this case, $L(A_s) = \{a\} \subseteq L(A'_s) = \{a, aa\}$.

Below we describe the construction [117] for PREREPLACE(DFA A_1, DFA A_2, DFA A_3). The steps are similar to the construction for POSTREPLACE. We first insert marks to identify all sub strings of strings in $\mathcal{L}(A_1)$. We construct A'_1, s.t.

$$\mathcal{L}(A'_1) = \{w' \mid k > 0, w = w_1 x_1 w_2 \ldots w_k x_k w_{k+1} \in \mathcal{L}(A_1),$$
$$w' = w_1 \natural_1 x_1 \natural_2 w_2 \ldots w_k \natural_1 x_k \natural_2 w_{k+1}\}.$$

Next, we construct A'_2, s.t.

$$\mathcal{L}(A'_2) = \{w' \mid k > 0, w' = w_1 \natural_1 x_1 \natural_2 w_2 \ldots w_k \natural_1 x_k \natural_2 w_{k+1},$$
$$\forall 1 \leq i \leq k, x_i \in \mathcal{L}(A_3), \forall 1 \leq i \leq k+1, w_i \in \mathcal{L}(A_h)\},$$

where $\mathcal{L}(A_h)$ is the set of strings which do not contain any substring in $\mathcal{L}(A_2)$. The language $\mathcal{L}(A_h)$ is defined as the complement set of $\{w_1 x w_2 \mid x \in \mathcal{L}(A_2), w_1, w_2 \in \Sigma^*\}$. Let A^* be the DFA accepting Σ^*. A_h can be constructed by taking the complement of POSTCONCAT(POSTCONCAT(A^*, A_2), A^*). Note that here we still use A_2 to construct A_h. The intuition is to identify the parts that remain the same (the substrings that are not in $\mathcal{L}(A_2)$) in the post image computation. We take intersection of A'_1 and A'_2 to mark matches (in A_3) and replace all these matches with A_2 to construct the final automata similar to the construction presented for POSTREPLACE.

When A_3 accepts an empty string, POSTREPLACE(A_1, A_2, A_3) performs *deletion*. I.e., it will delete all the matches in $\mathcal{L}(A_1)$. In this case, to compute the pre-image of the target strings, we would not be able to find any match of A_3 (an empty string in this case) to be replaced with A_2. For instance, if we compute PREREPLACE(DFA A_1, DFA A_2, DFA A_3) with POSTREPLACE(A_1, A_3, $A_3 \sqcup A_2$) when A_3 accepts only an empty string, it will return A_1 without any changes. To deal with deletion, we conservatively generate A for PREREPLACE(DFA A_1, DFA A_2, DFA A_3) that accepts $\mathcal{L}(A_2)$ to be repeated arbitrary times between any character of $\mathcal{L}(A_1)$. Formally speaking, A accepts the language $\{w_0 c_0 \ldots w_n c_n w_{n+1} \mid c_0 \ldots c_n \in \mathcal{L}(A_1), \forall_i, w_i \in \mathcal{L}(A_2^*)\}$, where $\mathcal{L}(A_2^*)$ denotes the closure of $\mathcal{L}(A_2)$.

4.6 Summary

In this chapter we discussed the use of automata as a symbolic representation for analyzing string manipulating programs. Automata represent sets of strings and support the operations needed for symbolic reachability analysis. Hence, automata are a suitable symbolic representation for string analysis. We discussed symbolic representation of automata where the transition relation of the automata are encoded as Multi-terminal Binary Decision Diagrams (MTBDD). We discussed the post- and pre-condition computations using automata as a symbolic representations, focusing on the concatenation and replacement operations. In the following chapters we discuss string analysis techniques that use automata as a symbolic representation.

Chapter 5
Relational String Analysis

The string analysis techniques we discussed in the previous chapter are not relational, i.e., they cannot keep track of relationships among variables. An analysis technique that is able to keep track of the relationships among the string variables, can improve the precision of the string analysis and enable verification of assertions such as $X_1 = X_2$ where X_1 and X_2 are string variables. It is not possible to prove such assertions using the techniques described in the previous chapter unless the set of values for string variables are singleton sets.

5.1 Relations Among String Variables

Sometimes it is essential to keep relations among string variables to prove assertions. Consider a simple branch statement in Fig. 5.1. Previous automata-based string analysis techniques that use single-track automata are not able to prove the assertion at the end of this program segment. Consider a symbolic analysis technique that uses one automaton for each variable at each program point to represent the set of values that the variables can take at that program point. Using this symbolic representation we can do a forward fixpoint computation to compute the reachable state space of the program. For example, the automaton for variable X_1 at the beginning of statement 2, call it $A_{X_1,2}$, will recognize the set $\mathcal{L}(A_{X_1,2}) = \Sigma^*$ to indicate that the input can be any string. Similarly, the automaton for variable X_2 at the beginning of statement 3, call it $A_{X_2,3}$, will recognize the set $\mathcal{L}(A_{X_2,3}) = \Sigma^*$. The question is how to handle the branch condition in statement 3. If we are using single track automata, all we can do at the beginning of statement 6 is the following: $\mathcal{L}(A_{X_1,6}) = \mathcal{L}(A_{X_2,6}) = \mathcal{L}(A_{X_1,3}) \cap \mathcal{L}(A_{X_2,3})$. The situation with the else branch is even worse. All we can do at line 4 is to set $\mathcal{L}(A_{X_1,4}) = \mathcal{L}(A_{X_1,3})$ and $\mathcal{L}(A_{X_2,4}) = \mathcal{L}(A_{X_2,3})$. Both branches will result in $\mathcal{L}(A_{X_1,7}) = \Sigma^*.c$ and $\mathcal{L}(A_{X_2,7}) = \Sigma^*$, which is clearly not strong enough to prove the assertion.

© Springer International Publishing AG 2017
T. Bultan et al., *String Analysis for Software Verification and Security*,
https://doi.org/10.1007/978-3-319-68670-7_5

Fig. 5.1 An example with a
branch statement

```
1: input X1;
2: input X2;
3: if (X1 = X2) goto 6;
4: X1 := X2 . c;
5: goto 7;
6: X1 := X1 . c;
7: assert (X1 = X2 . c);
```

Fig. 5.2 An example with a
loop

```
1: X1 := a;
2: X2 := a;
3: X1 := X1 . b;
4: X2 := X2 . b;
5: assert (X1 = X2);
6: goto 3;
```

Relational string analysis can be used to verify the assertion in the above program via modeling the relations among string variables. In the following, we use a single multi-track automaton for each program point, where each track of the automaton corresponds to one string variable. For the above example, the multi-track automaton at the beginning of statement 3 will accept any pairs of strings X_1, X_2 where $X_1, X_2 \in \Sigma^*$. However, the multi-track automaton at the beginning of statement 6 will only accept pairs of strings X_1, X_2 where $X_1, X_2 \in \Sigma^*$ and $X_1 = X_2$. We compute the post-condition $(\exists X_1.(X_1 = X_2) \wedge (X_1' = X_1.c))[X_1/X_1']$ and generate the multi-track automaton that only accepts pairs of strings X_1, X_2 where $X_1, X_2 \in \Sigma^*$ and $X_1 = X_2.c$. Similarly, the multi-track automaton at the beginning of statement 4 will only accept pairs of strings X_1, X_2 where $X_1, X_2 \in \Sigma^*$ and $X_1 \neq X_2$, and after the assignment, we will generate the multi-track automaton that only accepts pairs of strings X_1, X_2 where $X_1, X_2 \in \Sigma^*$ and $X_1 = X_2.c$. Hence, we are able to prove the assertion in statement 7.

Now, consider another string manipulation segment shown in Fig. 5.2, which contains an infinite loop. If we try to compute the reachable states of this program by iteratively adding states that can be reached after a single step of execution, our analysis will never terminate. We incorporate a widening operator to accelerate our symbolic reachability computation and compute an over-approximation of the fixpoint that characterizes the reachable states. Note that the assertion in this program segment is not explicitly established, i.e., there is no assignment, such as $X_1 := X_2$, or branch condition, such as $X_1 = X_2$, that implies that this assertion holds. Also, the assertion specifies the equality among two string variables. Analysis techniques that encode reachable states using multiple single-track DFAs will raise a false alarm, since, individually, X_1 can be *abb* and X_2 can be *ab* at program point 5, but they cannot take these values at the same time. It is not possible to express such a constraint using single-track automata.

For this example, the multi-track automata based string analysis technique we discuss in this Chapter terminates in three iterations and computes the precise result. The multi-track automaton that characterizes the values of string variables X_1 and X_2 at program point 5, call it A_5, recognizes the language: $\mathcal{L}(A_5) = (a, a)(b, b)^+$.

Since $\mathcal{L}(A_5) \subseteq \mathcal{L}(X_1 = X_2)$, we conclude that the assertion holds. Although in this case the result of our analysis is precise, it is not guaranteed to be precise in general. However, it is guaranteed to be an over-approximation of the reachable states. Hence, our analysis is sound and if we conclude that an assertion holds, the assertion is guaranteed to hold for every program execution.

5.2 Multi-Track DFAs

We use multi-track DFAs to encode the relations among string variables, where each track represents the value of a string variable.

A multi-track DFA is a DFA with an alphabet that consists of many tracks. An n-track alphabet is defined as $(\Sigma \cup \{\lambda\})^n$, where $\lambda \notin \Sigma$ is a special symbol for padding. We use $w[i]$ ($1 \leq i \leq n$) to denote the ith track of $w \in (\Sigma \cup \{\lambda\})^n$. An *aligned* multi-track DFA is a multi-track DFA where all tracks are left justified (i.e., λ's are right justified). That is, if w is accepted by an aligned n-track DFA A, then for $1 \leq i \leq n, w[i] \in \Sigma^*\lambda^*$. We also use $\hat{w}[i] \in \Sigma^*$ to denote the longest λ-free prefix of $w[i]$. Aligned multi-track DFA languages are closed under intersection, union, and homomorphism. Let A_u be the aligned n-track DFA that accepts the (aligned) universe, i.e., $\{w \mid \forall i.w[i] \in \Sigma^*\lambda^*\}$. The complement of the language accepted by an aligned n-track DFA A is defined by *complement modulo alignment*, i.e., the intersection of the complement of $\mathcal{L}(A)$ with $\mathcal{L}(A_u)$. For the following descriptions, a multi-track DFA is an aligned multi-track DFA unless we explicitly state otherwise.

Consider the condition $X_1 = X_2.txt$ where X_1 is the same as X_2 concatenated with a constant string "txt". Figure 5.3 shows the multi-track DFA that models this constraint. It accepts the set of two-track strings (the first track is the value of X_1 and the second track is the value of X_2) such that the first track has the same value as the second track (staying at the state 0) until the second track hits and keeps the rest as the padding symbol λ (moving from state 0 to 1) to extend the first track to a constant string "txt" (and to state 2 and 3).

5.3 Relational String Analysis

The relational string analysis we present in this Chapter is based on the following observations: (1) The transitions and the states of a string manipulating program can be symbolically represented using word equations with existential quantification, (2) word equations can be represented/approximated using multi-track DFAs, which

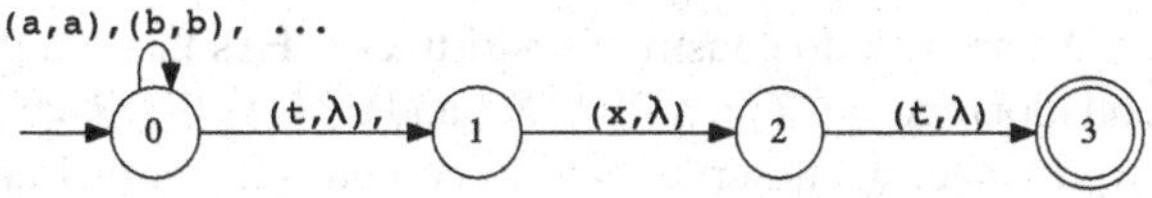

Fig. 5.3 A multi-track DFA that recognizes $X_1 = X_2.txt$

are closed under intersection, union, complement, and projection, and (3) the operations required during reachability analysis (such as equivalence checking) can be computed on DFAs.

Before we discuss how to perform relational symbolic reachability analysis for string manipulating programs, we introduce the word equations in this section. We characterize word equations that can be expressed using multi-track DFAs, as well as discuss the construction of these multi-track DFAs. Using these constructions, in the next section, we show how to perform symbolic reachability analysis for string manipulating programs. We iteratively compute post images of reachable states represented as multi-track DFAs and join the results using automata widening until reaching a fixpoint.

5.3.1   Word Equations

A word equation is an equality relation of two words that concatenate variables from a finite set $\mathbf{X}$ and words from a finite set of constants $\mathcal{C}$. The general form of word equations is defined as $v_1 \ldots v_n = v'_1 \ldots v'_m$, where $\forall i, v_i, v'_i \in \mathbf{X} \cup \mathcal{C}$.

As shown in [134], word equations and Boolean combinations ($\neg$, $\wedge$ and $\vee$) of these equations can be expressed using equations of the form $X_1 = X_2 c$, $X_1 = cX_2$, $c = X_1 X_2$, $X_1 = X_2 X_3$, Boolean combinations of such equations and existential quantification. For example, a word equation $f : X_1 = X_2 dX_3 X_4$ is equivalent to $\exists X_{k_1}, X_{k_2}.X_1 = X_2 X_{k_1} \wedge X_{k_1} = dX_{k_2} \wedge X_{k_2} = X_3 X_4$.

Let f be a word equation over $\mathbf{X} = \{X_1, X_2, \ldots, X_n\}$ and $f[c/X]$ denote a new equation where X is replaced with c for all X that appears in f. We say that an n-track DFA A *under-approximates* f if for all $w \in \mathcal{L}(A)$, $f[\hat{w}[1]/X_1, \ldots, \hat{w}[n]/X_n]$ holds. We say that an n-track DFA A *over-approximates* f if for any $s_1, \ldots, s_n \in \Sigma^*$ where $f[s_1/X_1, \ldots, s_n/X_n]$ holds, there exists $w \in \mathcal{L}(A)$ such that for all $1 \leq i \leq n$, $\hat{w}[i] = s_i$. We call A *precise* with respect to f if A both under-approximates and over-approximates f. A word equation f is regularly expressible if and only if there exists a multi-track DFA A such that A is precise with respect to f.

We will discuss how to construct the corresponding multi-track DFAs for the basic forms of word equations: (1) $X_1 = X_2 c$, (2) $X_1 = cX_2$, (3) $c = X_1 X_2$, and (4) $X_1 = X_2 X_3$.

Following have been shown in [134].

1. $X_1 = X_2 c$, $X_1 = cX_2$, and $c = X_1 X_2$ are regularly expressible, as well as their Boolean combinations.
2. $X_1 = cX_2$ is regularly expressible but the corresponding A has exponential number of states in the length of c.
3. $X_1 = X_2 X_3$ is not regularly expressible.

We are able to construct multi-track DFAs that are precise with respect to word equations: $X_1 = X_2 c$, $X_1 = cX_2$, and $c = X_1 X_2$. Since $X_1 = X_2 X_3$ is not regularly expressible, we describe how to compute DFAs that approximate such non-linear word equations. Using the DFA constructions for these four basic forms we can construct multi-track DFAs for all word equations and their Boolean combinations (if the word equation contains a non-linear term then the constructed DFA will approximate the equation, otherwise it will be precise). The Boolean operations conjunction, disjunction and negation can be handled with intersection, union, and complement modulo alignment of the multi-track DFAs, respectively. Existential quantification on the other hand, can be handled using homomorphism, where given a word equation f and a multi-track automaton A such that A is precise with respect to f, then the multi-track automaton $A \downarrow_i$ is precise with respect to $\exists X_i.f$ where $A \downarrow_i$ denotes the result of erasing the ith track (by homomorphism) of A.

5.4 Construction of Multi-Track DFAs for Basic Word Equations

Given a DFA $A = \langle Q, \Sigma, \delta, I, F \rangle$, Q is the set of states, Σ is the alphabet, $\delta : Q \times \Sigma \rightarrow Q$ is the transition function, $I \in Q$ is the initial state, and $F \subseteq Q$ is the set of final (accepting) states. We say a state $q \in Q$ is a *sink* state if $q \notin F$ and $\forall a \in \Sigma, \delta(q, a) = q$. In the following construction functions, we ignore transitions that go to sink states, and assume that all unspecified transitions go to sink states.

Before we give the construction functions, we generalize the problem of constructing multi-track DFAs for word equations as follows. We assume that each variable in $\mathbf{X} = \{X_1, X_2, \ldots, X_n\}$ is associated with an automaton $A_i = \langle Q_i, \Sigma, \delta_i, I_i, F_i \rangle$, where $\mathcal{L}(A_i)$ denotes the set of values that the variable X_i can take. Then, given a word equation f over $\mathbf{X}$, we say that *an n-track DFA A under-approximates f within* $A_1, \ldots A_n$, if for all $w \in \mathcal{L}(A), f[\hat{w}[1]/X_1, \ldots, \hat{w}[n]/X_n]$ holds and for all $1 \leq i \leq n, \hat{w}[i] \in \mathcal{L}(A_i)$. We say that *an n-track DFA A over-approximates f within* $A_1, \ldots A_n$, if for any $s_1, \ldots, s_n \in \Sigma^*$ where $f[s_1/X_1, \ldots, s_n/X_n]$ holds and for all $1 \leq i \leq n, s_i \in \mathcal{L}(A_i)$, there exists $w \in \mathcal{L}(A)$ such that for all $1 \leq i \leq n, \hat{w}[i] = s_i$. Note that, for either case, for any word $w \in \mathcal{L}(A)$, for all $1 \leq i \leq n, \hat{w}[i] \in \mathcal{L}(A_i)$.

Below we define the construction function $\mathcal{A}$ that returns the corresponding automata for each basic word equation.

Construction of $\mathcal{A}(X_1 = X_2 c, A_1, A_2)$

Let $A_1 = \langle Q_1, \Sigma, \delta_1, I_1, F_1 \rangle$, $A_2 = \langle Q_2, \Sigma, \delta_2, I_2, F_2 \rangle$ be two DFAs that accept possible values of variables X_1 and X_2, respectively. $\mathcal{A}(X_1 = X_2 c, A_1, A_2)$ defines the construction function of a 2-track DFA $A = \langle Q, \Sigma, \delta, I, F \rangle$, such that A is precise with respect to $X_1 = X_2 c$ within A_1, A_2.

Let $c = a_1a_2 \ldots a_n$, where $\forall 1 \leq i \leq n, a_i \in \Sigma$ and n is the length of the constant string c. $\mathcal{A}(X_1 = X_2c, A_1, A_2)$ returns $A = \langle Q, (\Sigma \cup \{\lambda\})^2, \delta, q_0, F \rangle$, which is constructed as:

- $Q \subseteq Q_1 \times Q_2 \times \{0, \ldots, n\}$,
- $I = (I_1, I_2, 0)$,
- $\forall a \in \Sigma, \delta((r, p, 0), (a, a)) = (\delta_1(r, a), \delta_2(p, a), 0)$,
- $\forall a_i, p \in F_2, \delta((r, p, i), (a_i, \lambda)) = (\delta_1(r, a_i), p, i + 1)$,
- $F = \{(r, p, i) \mid r \in F_1, p \in F_2, i = n\}$.

Note that A simulates A_1 and A_2 making sure that both tracks are the same until a final state of A_2 is reached. Then, the second track reads the symbol λ while the first track reads the constant c, and the automaton goes to a final state when c is consumed. $|Q|$ is $O(|Q_1| \times |Q_2| + n)$ since in the worst case Q will contain all possible combinations of states in Q_1 and Q_2 followed with a tail of n states for recognizing the constant c. For the automaton A resulting from the above construction we have, $w \in \mathcal{L}(A)$ if and only if $\hat{w}[1] = \hat{w}[2]c$, $\hat{w}[1] \in \mathcal{L}(A_1)$ and $\hat{w}[2] \in \mathcal{L}(A_2)$, i.e., A is precise with respect to $X_1 = X_2c$ (within A_1, A_2), and, hence, $X_1 = X_2c$ is regularly expressible.

Construction of $\mathcal{A}(X_1 = cX_2, A_1, A_2)$

Let $A_1 = \langle Q_1, \Sigma, \delta_1, I_1, F_1 \rangle$, $A_2 = \langle Q_2, \Sigma, \delta_2, I_2, F_2 \rangle$ be two DFAs that accept possible values of variables X_1 and X_2, respectively. Below we present $\mathcal{A}(X_1 = cX_2, A_1, A_2)$, the construction function that returns a 2-track DFA A, such that A is precise with respect to $X_1 = cX_2$ within A_1, A_2. Let $c = a_1a_2 \ldots a_n$, where $\forall 1 \leq i \leq n, a_i \in \Sigma$ and n is the length of the constant string c.

The intuition behind the construction of A is as follows. In the initial stage (denoted as *init* below), A makes sure that the first track matches the constant c, while recording the string that is read in the second track in a buffer (a vector of symbols) stored in its state. After c is consumed, A goes to the next stage (denoted as *match* below) and matches the symbols read in the first track with the next symbol stored in the buffer while continuing to store the symbols read in the second track in the buffer. Note that, the kth symbol read in track 2 has to be matched with the $(k + n)$th symbol read in track 1. So, the buffer stores the symbols read in track 2 until the corresponding symbol in track 1 is observed.

Let $\vec{v}$ be a size n vector. For $1 \leq i \leq n, \vec{v}[i] \in \Sigma \cup \{\bot\}$. The vector $\vec{v}' = \vec{v}[i := a]$ is defined as follows: $\vec{v}'[i] = a$ and $\forall j \neq i, \vec{v}'[j] = \vec{v}[j]$. $A = \langle Q, (\Sigma \cup \{\lambda\})^2, \delta, I, F \rangle$ is constructed as:

- $Q \subseteq Q_1 \times Q_2 \times \{1, \ldots, n\} \times (\Sigma \cup \{\bot\})^n \times \{init, match\}$,
- $I = (I_1, I_2, 1, \vec{v}_\bot, init)$, where $\forall i, \vec{v}_\bot[i] = \bot$,
- $\forall a \in \Sigma, 1 \leq i < n, \delta((r, p, i, \vec{v}, init), (a_i, a)) = (\delta_1(r, a_i), \delta_2(p, a), i + 1, \vec{v}[i := a], init)$,

- $\forall a \in \Sigma, i = n, \delta((r, p, i, \vec{v}, init), (a_i, a)) = (\delta_1(r, a_i), \delta_2(p, a), 1, \vec{v}[i := a],$
  $match),$
- $\forall a, b \in \Sigma, 1 \leq i < n, \vec{v}[i] = a, \delta((r, p, i, \vec{v}, match), (a, b)) = (\delta_1(r, a), \delta(p, b),$
  $i + 1, \vec{v}[i := b], match),$
- $\forall a, b \in \Sigma, i = n, \vec{v}[i] = a, \delta((r, p, i, \vec{v}, match), (a, b)) = (\delta_1(r, a), \delta(p, b), 1,$
  $\vec{v}[i := b], match),$
- $\forall a \in \Sigma, p \in F_2, 1 \leq i < n, \vec{v}[i] = a, \delta((r, p, i, \vec{v}, match), (a, \lambda)) = (\delta_1(r, a),$
  $p, i + 1, \vec{v}[i := \perp], match),$
- $\forall a \in \Sigma, p \in F_2, i = n, \vec{v}[i] = a, \delta((r, p, i, \vec{v}, match), (a, \lambda)) = (\delta_1(r, a),$
  $p, 1, \vec{v}[i := \perp], match),$
- $F = \{(r, p, i, \vec{v}_\perp, match) \mid r \in F_1, p \in F_2\}.$

Since A accepts the set $\{w \mid \hat{w}[1] = c\hat{w}[2], \hat{w}[1] \in \mathcal{L}(A_1), \hat{w}[2] \in \mathcal{L}(A_2)\}$, $X_1 = cX_2$ is regularly expressible. However, the number of states of A is exponential in c. Below, we show that the exponential number of states is inevitable.

Intractability of $X_1 = cX_2$

Consider the equation $X_1 = cX_2$, where c is a constant string of length n. Let $\mathcal{L}(A_1)$ and $\mathcal{L}(A_2)$ be regular languages. Define the 2-track language:

$$L = \{(x_1x_2, y_1y_2\lambda^n) \mid x_1x_2 \in \mathcal{L}(A_1), y_1y_2 \in \mathcal{L}(A_2), k \geq n, |x_1x_2| = k, |x_1| =$$
$$|y_1| = n, x_1 = c, x_2 = y_1y_2\}$$

Note that any automaton A that accepts the language L defined above will be precise with respect to the the equation $X_1 = cX_2$ (within A_1 and A_2).

In fact, any nondeterministic finite automaton (NFA) A needs at least 2^n states to accept L. Let $c = 1^n$ and consider the regular languages $\mathcal{L}(A_1) = (0 + 1)^+$ and $\mathcal{L}(A_2) = (0 + 1)^+$. Suppose A is an NFA accepting L. Consider any pair of distinct strings y_1 and y_1' of length n. Then A will accept the following 2-track strings:

$(1^n x_2, y_1 y_2 \lambda^n)$, where $x_2, y_1, y_2 \in (0 + 1)^+, k \geq n, |1^n x_2| = k, |y_1| = n, x_2 = y_1 y_2$, and

$(1^n x_2', y_1' y_2' \lambda^n)$, where $x_2', y_1', y_2' \in (0+1)^+, k \geq n, |1^n x_2'| = k, |y_1'| = n, x_2' = y_1' y_2'$

Suppose in processing $(1^n x_2, y_1 y_2 \lambda^n)$, A enters state q after processing the initial 2-track segment $(1^n, y_1)$, and in processing $(1^n x_2', y_1' y_2' \lambda^n)$, A enters state q' after processing the initial 2-track segment $(1^n, y_1')$. Then $q \neq q'$; otherwise, A will also accept $(1^n x_2, y_1' y_2 \lambda^n)$. This is a contradiction, since $x_2 \neq y_1' y_2$.

Since there are 2^n distinct strings y of length n, it follows that A must have at least 2^n states.

Construction of $\mathcal{A}(c = X_1X_2, A_1, A_2)$

Below we briefly describe $\mathcal{A}(c = X_1X_2, A_1, A_2)$, the construction function that returns a 2-track DFA A, such that A is precise with respect to $c = X_1X_2$ within the given regular sets characterizing possible values of X_1 and X_2. Assume that $c = a_1 \ldots a_n$. We can split c to two strings $a_1 \ldots a_k$ and $a_{k+1} \ldots a_n$ so that $c = a_1 \ldots a_k a_{k+1} \ldots a_n$. There are $n+1$ such splits. For each of them, if $a_1 \ldots a_k \in \mathcal{L}(A_1)$ and $a_{k+1} \ldots a_n \in \mathcal{L}(A_2)$, then if $k \geq n - k$, $(a_1 \ldots a_k, a_{k+1} \ldots a_n \lambda^{2k-n})$ should be accepted by A and if $k < n - k$, $(a_1 \ldots a_k \lambda^{n-2k}, a_{k+1} \ldots a_n)$ should be accepted by A. We can construct an automaton A with $O(n^2)$ states that accepts this language by explicitly checking each of these $n + 1$ cases. Since we can construct this 2-track DFA, it follows that $c = X_1X_2$ is regularly expressible.

Construction of $\mathcal{A}(X_1 = X_2X_3, A_1, A_2, A_3)$

Since it has been shown in [134] that $X_1 = X_2X_3$ is not regularly expressible, $\mathcal{A}(X_1 = X_2X_3, A_1, A_2, A_3)$ defines a conservative construction that accepts (*over* or *under*) approximation of $X_1 = X_2X_3$. We first propose $\mathcal{A}^+(X_1 = X_2X_3, A_1, A_2, A_3)$ as an *over* approximation construction for $X_1 = X_2X_3$. Let $A_1 = \langle Q_1, \Sigma, \delta_1, I_1, F_1 \rangle$, $A_2 = \langle Q_2, \Sigma, \delta_2, I_2, F_2 \rangle$, and $A_3 = \langle Q_3, \Sigma, \delta_3, I_3, F_3 \rangle$ accept values of X_1, X_2, and X_3, respectively. $\mathcal{A}^+(X_1 = X_2X_3, A_1, A_2, A_3)$ returns the automata $A = \langle Q, (\Sigma \cup \{\lambda\})^3, \delta, I, F \rangle$, which is constructed as follows.

- $Q \subseteq Q_1 \times Q_2 \times Q_3 \times Q_3$,
- $I = (I_1, I_2, I_3, I_3)$,
- $\forall a, b \in \Sigma, \delta((r, p, s, s'), (a, a, b)) = (\delta_1(r, a), \delta_2(p, a), \delta_3(s, b), s')$,
- $\forall a, b \in \Sigma, p \in F_2, s \notin F_3, \delta((r, p, s, s'), (a, \lambda, b)) = (\delta_1(r, a), p, \delta_3(s, b), \delta_3(s', a))$,
- $\forall a \in \Sigma, p \in F_2, s \in F_3, \delta((r, p, s, s'), (a, \lambda, \lambda)) = (\delta_1(r, a), p, s, \delta_3(s', a))$,
- $\forall a \in \Sigma, p \notin F_2, s \in F_3, \delta((r, p, s, s'), (a, a, \lambda)) = (\delta_1(r, a), \delta_2(p, a), s, s')$,
- $F = \{(r, p, s, s') \mid r \in F_1, p \in F_2, s \in F_3, s' \in F_3\}$.

The intuition is as follows: $\mathcal{A}^+(X_1 = X_2X_3, A_1, A_2, A_3)$ returns the automata A that traces A_1, A_2 and A_3 on the first, second and third tracks, respectively, and makes sure that the first and second tracks match each other. After reaching an accepting state in A_2, A enforces the second track to be λ and starts to trace A_3 on the first track to ensure the rest (suffix) is accepted by A_3. $|Q|$ is $O(|Q_1| \times |Q_2| \times |Q_3| + |Q_1| \times |Q_3| \times |Q_3|)$. For all $w \in \mathcal{L}(A)$, the following hold:

- $\hat{w}[1] \in \mathcal{L}(A_1), \hat{w}[2] \in \mathcal{L}(A_2), \hat{w}[3] \in \mathcal{L}(A_3)$,
- $\hat{w}[1] = \hat{w}[2]w'$ and $w' \in \mathcal{L}(A_3)$,

Note that w' may not be equal to $\hat{w}[3]$, i.e., there exists $w \in \mathcal{L}(A)$, $\hat{w}[1] \neq \hat{w}[2]\hat{w}[3]$, and hence A is not precise with respect to $X_1 = X_2X_3$. On the other hand, for any w such that $\hat{w}[1] = \hat{w}[2]\hat{w}[3]$, we have $w \in \mathcal{L}(A)$, hence A is a regular *over*-approximation of $X_1 = X_2X_3$.

Below, we define $\mathcal{A}^-(X_1 = X_2X_3, A_1, A_2, A_3)$ for the conservative construction of a regular *under*-approximation of $X_1 = X_2X_3$ (which is necessary for conservative approximation of its complement set). We use the idea that if $\mathcal{L}(A_2)$ is a finite set language, one can construct the DFA A that satisfies $X_1 = X_2X_3$ by explicitly taking the union of the construction of $X_1 = cX_3$ (by calling $\mathcal{A}(X_1 = cX_3, A_1, A_3)$) for all $c \in \mathcal{L}(A_2)$. If $\mathcal{L}(A_2)$ is an infinite set language, we construct a regular *under*-approximation of $X_1 = X_2X_3$ by considering a (finite) subset of $\mathcal{L}(A_2)$ where the length is bounded. Formally speaking, for each $k \geq 0$ we can construct A_k, so that $w \in \mathcal{L}(A_k), \hat{w}[1] = \hat{w}[2]\hat{w}[3]$, $\hat{w}[1] \in \mathcal{L}(A_1)$, $\hat{w}[3] \in \mathcal{L}(A_3)$, $\hat{w}[2] \in \mathcal{L}(A_2)$ and $|\hat{w}[2]| \leq k$. It follows that A_k is a regular *under*-approximation of $X_1 = X_2X_3$. If $\mathcal{L}(A_2)$ is a finite set language, there exists k (the length of the longest accepted word) where $\mathcal{L}(A_k)$ is precise with respect to $X_1 = X_2X_3$. If $\mathcal{L}(A_2)$ is an infinite set language, there does not exist such k so that $\mathcal{L}(A_k)$ is precise with respect to $X_1 = X_2X_3$, as we have proven non-regularity of $X_1 = X_2X_3$.

In fact, for an infinite set language, we can always improve the precision by increasing k, i.e., for all k_1, there exists a k_2 such that $\mathcal{L}(A_{k_1}) \subset \mathcal{L}(A_{k_2})$ and $k_1 < k_2$.

While we cannot have precise construction of $X_1 = X_2X_3$, it would be of interest to define $\mathcal{A}^-(X_1 = X_2X_3, A_1, A_2, A_3)$ that returns the *tightest* regular *under*-approximation of $X_1 = X_2X_3$. We say a regular under-approximation A_κ is *tightest* if $\mathcal{L}(A_\kappa)$ is an under-approximation of $X_1 = X_2X_3$ and for all A' where A' is an under-approximation of $X_1 = X_2X_3$ we have $\mathcal{L}(A') \subseteq \mathcal{L}(A_\kappa)$. Since the precision of a regular under-approximation can be always improved by adding new words to the language, the tightest regular under-approximation does not exist if $\mathcal{L}(A_2)$ is not finite.

5.5 Symbolic Reachability Analysis

Based on the constructions we discussed above, we can further define a function $\mathcal{A}(exp:\text{word equation}, b:\text{bool})$ that takes a word equation as input and returns a corresponding multi-track DFA, if necessary either over-approximating (if $b = +$) or under-approximating (if $b = -$). We can use the $\mathcal{A}$ function to soundly approximate all word equations *and* their boolean combinations, including existentially-quantified word equations. The boolean operations conjunction, disjunction, and negation on word equations are handled using intersection, disjunction, and complement of the corresponding multi-track DFAs, respectively; existentially-quantified word equations are handled using homomorphisms (by projecting the track that corresponds to the quantified variable).

Given an assignment statement *stmt* of the form $X := exp$ we first represent it as a word equation of the form $X' = exp$ where *exp* is an expression on the current state variables, and X' denotes the next state variables. Then we abstract *stmt* by constructing a multi-track automaton A_{stmt} that over-approximates the corresponding word equation as follows $A_{stmt} = \mathcal{A}(X' = exp, +)$. A branch condition specified as an expression *exp* is similarly abstracted using $\mathcal{A}(X' = X \land exp, +)$ for the then branch and $\mathcal{A}(X' = X \land \neg exp, +)$ for the else branch. The result of the regular abstraction consists of the control flow graph of the original program where each statement in the control flow graph is associated with a multi-track DFA that over-approximates the behavior of the corresponding statement.

The relational symbolic reachability analysis consists of two phases. In the first phase, we use one multi-track DFA for each program point to symbolically represent possible values of string variables at that program point, where each track corresponds to one string variable. Our approach is based on a forward fixpoint computation on multi-track DFAs. We iteratively compute post-images of reachable states and join the results until we reach a fixpoint.

It is possible to extend symbolic reachability analysis to an inter-procedural analysis using function summaries. During the forward fixpoint computation if we encounter a call to a function that has not been summarized, we go to the second phase of the analysis, which is function summarization. Each function is summarized when needed, and once a function is summarized, the summary DFA is used to compute the return values at the call sites without going through the body of the function. During the summarization phase, (recursive) functions are summarized as unaligned multi-track DFAs that specify the relations among their inputs and return values. We first build (cyclic) dependency graphs to specify how the inputs flow to the return values. Each node in the dependency graph is associated with an unaligned multi-track DFA that traces the relation among inputs and the value of that node. We iteratively compute post images of reachable relations and join the results until we reach a fixpoint. Upon termination, the summary is the union of the unaligned DFAs associated with the return nodes. To compose these summaries at the call site, an alignment algorithm has been proposed in [134] to align (so that λ's are right justified) an unaligned multi-track DFA.

This reachability analysis is sound but incomplete due to the following approximations: (1) regular approximation for non-linear word equations, (2) the widening operation and (3) function summarization.

5.5.1 Forward Fixpoint Computation

The first phase of the analysis is a standard forward fixpoint computation on multi-track DFAs. Each program point is associated with a single multi-track DFA, where each track is associated with a single string variable $X \in \mathbf{X}$. We use $A[l]$ to denote the multi-track automaton at the program label l. The forward fixpoint computation algorithm (Algorithm 1) is a standard work-queue algorithm.

Algorithm 1 RELATIONALFORWARDRECAHABILITY(l_0)

1: Init(A);
2: queue WQ;
3: WQ.enqueue($l_0 : stmt_0$);
4: **while** $WQ \neq NULL$ **do**
5:     $e := WQ$.dequeue(); Let e be $l : stmt$;
6:     **if** *stmt* is seqstmt **then**
7:         $m :=$ POST($A[l]$, *stmt*);
8:         PROPAGATE($m, l + 1$);
9:     **end if**
10:     **if** *stmt* is if *exp* goto l' **then**
11:         $m := A[l] \sqcap \mathcal{A}(exp, +)$;
12:         **if** $\mathcal{L}(m) \neq \emptyset$ **then**
13:             PROPAGATE(m, l');
14:         **end if**
15:         $m := A[l] \sqcap \mathcal{A}(\neg exp, +)$;
16:         **if** $\mathcal{L}(m) \neq \emptyset$ **then**
17:             PROPAGATE($m, l + 1$);
18:         **end if**
19:     **end if**
20:     **if** *stmt* is assert *exp* **then**
21:         $m := \mathcal{A}(exp, -)$;
22:         **if** $\mathcal{L}(A[l]) \not\sqsubseteq \mathcal{L}(m)$ **then**
23:             ASSERTFAILED(l);
24:         **else**
25:             PROPAGATE($A[l], l + 1$);
26:         **end if**
27:     **end if**
28:     **if** *stmt* is goto L **then**
29:         **for** $l' \in L$ **do**
30:             PROPAGATE($A[l], l'$);
31:         **end for**
32:     **end if**
33: **end while**

Algorithm 2 PROPAGATE(m, l)

1: $m' := (m \sqcup A[l]) \nabla A[l]$;
2: **if** $m' \not\sqsubseteq A[l]$ **then**
3:     $A[l] := m'$;
4:     WQ.enqueue(l);
5: **end if**

Initially, for all labels l, $\mathcal{L}(A[l]) = \emptyset$. We iteratively compute the post-images of the statements and join the results to the corresponding automata. For a *stmt* in the form: $X := sexp$, the post-image is computed as:

$$\text{POST}(A, stmt) \equiv (\exists X.A \sqcap \mathcal{A}(X' = sexp, +))[X/X'].$$

$\mathcal{A}(exp, b)$ calls the corresponding construction function, e.g., $\mathcal{A}^+(X_1 = X_2X_3, A_1,$ $A_2, A_3)$, to return the DFA that accepts a regular approximation of *exp*, where $b \in \{+, -\}$ indicates the direction (*over* or *under*, respectively) of approximation if needed. During the construction, we recursively push the negations ($\neg$) (and flip the direction) inside to the basic expressions (*bexp*), and use the corresponding construction of multi-track DFAs discussed in the previous section.

During the fixpoint computation, we report assertion failures if $A[l]$ accepts some string that violates the assertion labeled l. Note that at line 21 we compute an under approximation of the assertion expression to ensure the soundness of our analysis. Finally, a program label l is not reachable if $\mathcal{L}(A[l])$ is empty.

5.5.2 Summarization

We use function summaries to handle function calls. Each function f is summarized as a finite state transducer, denoted as A_f, which captures the relations among input variables (parameters), denoted as X_p, and return values. The return values are tracked in the output track, denoted as X_o. We discuss the generation of the transducer A_f below. For a *stmt* in the form $X := \mathtt{call}\, f(e_1, \ldots, e_n)$, the post-image is computed as:

$$\text{POST}(A, stmt) \equiv (\exists X, X_{p_1}, \ldots X_{p_n} . A \sqcap A_I \sqcap A_f)[X/X_o],$$

where $A_I = \mathcal{A}(\bigwedge_i X_{p_i} = e_i, +)$. The process terminates when we reach a fixpoint. To accelerate the fixpoint computation, we extend automata widening operators (discussed in Sect. 6.5), denoted as ∇, to multi-track automata. We identify equivalence classes according to specific equivalence conditions and merge states in the same equivalence class [17, 23]. The equality relations among tracks are preserved while widening multi-track automata. That is, if $\mathcal{L}(A) \subseteq \mathcal{L}(x = y)$ and $\mathcal{L}(A') \subseteq \mathcal{L}(x = y)$, $\mathcal{L}(A \nabla A') \subseteq \mathcal{L}(x = y)$.

5.6 Summary

In this Chapter, we discussed the relational string verification techniques based on multi-track automata. The presented approach is capable of verifying properties that depend on relations among string variables. We discussed the basic word equations (over string concatenations) and the corresponding automata constructions. The presented verification technique is based on forward symbolic reachability analysis with multi-track automata, conservative approximations of word equations and function summarization.

Chapter 6
Abstraction and Approximation

Verifying string manipulating programs is a crucial problem in computer security. String operations are used extensively within web applications to manipulate user input, and their erroneous use is the most common cause of security vulnerabilities in web applications. Unfortunately, verifying string manipulating programs is an undecidable problem in general and any approximate string analysis technique has an inherent tension between efficiency and precision. In this Chapter we present a set of sound abstractions for strings and string operations that allow for both efficient and precise verification of string manipulating programs. Particularly, we are able to verify properties that involve implicit relations among string variables. We first describe an abstraction called regular abstraction which enables us to perform string analysis using multi-track automata as a symbolic representation. We then introduce two other abstractions—alphabet abstraction and relation abstraction—that can be used in combination to tune the analysis precision and efficiency. We show that the relation and alphabet abstractions can be composed with the regular abstraction (and with each other) to obtain a family of abstractions. In fact, these abstractions form an abstraction lattice that generalizes the string analysis techniques studied previously in isolation, such as size analysis or non-relational string analysis.

6.1 Regular Abstraction

The regular abstraction maps a set of string tuples to a set of string tuples accepted by a multi-track automaton. This enables us to use deterministic finite state automata (DFAs) as a symbolic representation during string analysis. As we discussed earlier, a *multi-track automaton* (or multi-track DFA) is a DFA that has transitions on tuples of characters rather than single characters. For a given alphabet Σ let $\Sigma_\lambda = \Sigma \cup \{\lambda\}$, where $\lambda \notin \Sigma$ is a special padding character. An n-track alphabet is defined as $\Sigma^n = \Sigma_\lambda \times \cdots \times \Sigma_\lambda$ (n times). A track corresponds to a particular position in the

© Springer International Publishing AG 2017

T. Bultan et al., *String Analysis for Software Verification and Security*,
https://doi.org/10.1007/978-3-319-68670-7_6

n-tuple. A multi-track DFA is *aligned* if and only if for all words w accepted by the DFA, $w \in \Sigma^* \lambda^*$ (i.e., all padding is at the end of the word). Using aligned multi-track automata gives us a representation that is closed under intersection and can be converted to a canonical form after determinization and minimization. In the following, multi-track DFAs are assumed to be aligned unless explicitly stated otherwise.

As we discussed earlier, the statements of a string manipulating program can be represented as word equations. A word equation is an equality relation between two terms, each of which is a finite concatenation of string variables and string constants. Regular abstraction abstracts a given program by mapping the word-equations representing the program statements to multi-track DFAs. Since word equations can not be precisely represented using multi-track automata, we use the results presented in the previous chapter to construct a sound abstraction of the given program (i.e., in the abstracted program the set of values that a variable can take is a superset of the possible values that a variable can take in the concrete program). Note that, since branch conditions can contain negated terms, we need to be able to construct both an over- and an under-approximation of a given word equation. We construct multi-track automata that precisely represent word equations when possible, and either over- or under-approximate the word equations (as desired) otherwise.

The abstract domain that results from the regular abstraction is defined as a lattice on multi-track automata over an alphabet Σ^n. We denote this automata lattice as $\mathcal{L}_A = (\overline{A_{\Sigma^n}}, \sqsubseteq, \sqcup, \sqcap, \perp, \top)$, where $\overline{A_{\Sigma^n}}$ is the set of multi-track automata over the alphabet Σ^n. For $A_1, A_2 \in \overline{A_{\Sigma^n}}$, $A_1 \sqsubseteq A_2$ if and only if $\mathcal{L}(A_1) \subseteq \mathcal{L}(A_2)$. The bottom element is defined as $\mathcal{L}(\perp) = \emptyset$ and the top element is defined as $\mathcal{L}(\top) = (\Sigma^n)^*$. There may be multiple automata that accept the same language; the lattice treats these automata as equivalent. If we use minimized DFAs then there is a unique automaton for each point in the lattice up to isomorphism. All of the multi-track automata in this lattice are aligned [133] and hence all operations take aligned automata as input and return aligned automata as output.

Since the family of regular languages is not closed under infinite union, we can use the widening operator from [17] as the join operator where $A_1 \sqcup A_2 = A_1 \nabla A_2$. The meet operator can be defined from the join operator using language complement: let $\neg A$ denote an automaton such that $\mathcal{L}(\neg A) = \mathcal{L}(\top) - \mathcal{L}(A)$; then $A_1 \sqcap A_2 = \neg(\neg A_1 \nabla \neg A_2)$.

Note that a similar automata lattice can also be defined for single-track automata over a single-track alphabet Σ where $\mathcal{L}_A = (\overline{A_\Sigma}, \sqsubseteq, \sqcup, \sqcap, \perp, \top)$ as we discussed in Chap. 4.

6.2   Alphabet Abstraction

This abstraction targets the values taken on by string variables, mapping multiple alphabet symbols to a single abstract symbol. For example, consider a string variable that can take the value $\{ab, abc\}$ from the alphabet $\Sigma = \{a, b, c\}$. Abstracting the

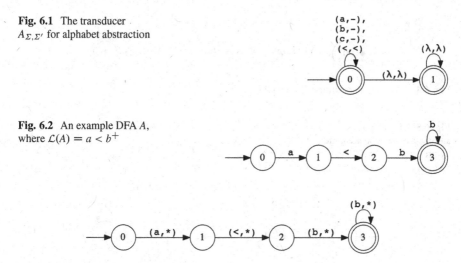

Fig. 6.1 The transducer $A_{\Sigma,\Sigma'}$ for alphabet abstraction

Fig. 6.2 An example DFA A, where $\mathcal{L}(A) = a < b^+$

Fig. 6.3 Extend A to a two-track automata A^2

symbols b and c yields the value $\{a-, a - -\}$, where $-$ stands for both b and c. The concretization of this abstract value would yield the value $\{ab, ac, abc, abb, acb, acc\}$. At the extreme, this abstraction can abstract out all alphabets symbols (in the above example, this would yield the abstract value $\{--, - - -\}$). In this case the only information retained from the original value is the length of the strings; all information about the content of the strings is lost. This abstraction is still useful in checking properties related to string length—we will return to this point in Sect. 6.4.

Before giving formal definitions, let us start from a simple example. Consider the concrete alphabet $\Sigma = \{a, b, c, <\}$ and the abstract alphabet $\Sigma' = \{<, -\}$, where we intended to abstract away all symbols as $-$ except $<$. We firs construct a two track automata (also known as transducer) that maps the input track (concrete values) to the output track (abstract values). In this case, we build $A_{\Sigma,\Sigma'}$ as shown in Fig. 6.1. The transducer can then be used to abstract concrete string values of an automaton to its abstract string values (abstraction), or to concretize abstract string values of an automaton to its concrete string values (concretization). Consider an example DFA A that accepts $a < b^+$ as show in Fig. 6.2. Below we show how to conduct abstraction and concretization with the transducer shown in Fig. 6.1.

The abstraction step is to construct an automata A_α that accepts the abstract strings of A. To do so, we first extend A to a two track DFA A^2 that has strings in the first track in $\mathcal{L}(A)$ but arbitrary strings in the second track (Fig. 6.3). In these figures, we use "*" to denote any alphabet symbol. Then, we take the intersection of A^2 and $A_{\Sigma,\Sigma'}$ to restrict the string values in the second track to the abstract strings of the values in the first track. This step for our example results in A_α^2 (Fig. 6.4). As the final step of abstraction, we simply project away the first track of A_α^2, and the result is the automata A_α, where $\mathcal{L}(A_\alpha) = - < -^+$, which accepts the set of abstract strings of A (Fig. 6.5).

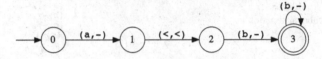

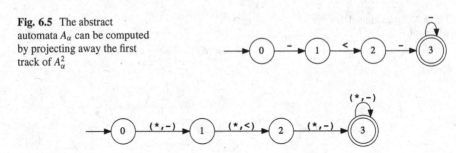

Fig. 6.4 Take the intersection of A^2 and $A_{\Sigma, \Sigma'}$ as A_α^2

Fig. 6.5 The abstract automata A_α can be computed by projecting away the first track of A_α^2

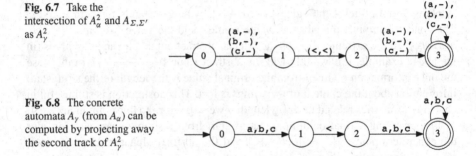

Fig. 6.6 Extend the abstract automata A_α a two-track automata A_α^2

Fig. 6.7 Take the intersection of A_α^2 and $A_{\Sigma, \Sigma'}$ as A_γ^2

Fig. 6.8 The concrete automata A_γ (from A_α) can be computed by projecting away the second track of A_γ^2

The concretization step is to construct an automata that accepts the concrete string values of the language of an abstract automata. To construct an automata A_γ that accepts the concrete strings of an abstract automata A_α (Fig. 6.5), we first extend A_α to a two track DFA A_α^2 (Fig. 6.6) that has strings in the second track in $\mathcal{L}(A_\alpha)$ but arbitrary strings in the first track. Similar to the abstraction step, we take the intersection of A_α^2 and $A_{\Sigma, \Sigma'}$ to restrict the string values in the first track corresponding the concrete strings of the values in the second track. In our example, this step results in A_γ^2 (Fig. 6.7), where the concrete string of the abstract string in the second track is now in the first track. As the final step, we simply project away the second track of A_γ^2, and the result is the automata A_γ (Fig. 6.8), where $\mathcal{L}(A_\gamma) = (a|b|c) < (a|b|c)^+$, which accepts all the concrete strings that can be mapped to the abstract strings in $\mathcal{L}(A_\alpha)$ (Fig. 6.5). Note that $\mathcal{L}(A_\gamma)$ is a super set of $\mathcal{L}(A)$ (Fig. 6.2).

Let us formally define the alphabet abstraction. The alphabet abstraction is parameterized by the choice of which symbols to abstract, hence it forms a family of abstractions. This family forms an abstraction lattice $\mathcal{L}_\Sigma$ called the *alphabet lattice* (distinct from the automata lattice introduced earlier). Let Σ, a finite alphabet, be the concrete alphabet, and $- \notin \Sigma$ be a special symbol to represent characters that are abstracted away. An abstract alphabet of Σ is defined as $\Sigma' \cup \{-\}$, where $\Sigma' \subseteq \Sigma$. The abstract alphabets of Σ form a complete lattice $\mathcal{L}_\Sigma = (\mathcal{P}(\Sigma \cup \{-\}), \sqsubseteq_\Sigma , \cup, \cap, \sigma_\perp, \sigma_\top)$ where the bottom element $\sigma_\perp$ is $\Sigma \cup \{-\}$, the top element $\sigma_\top$ is $\{-\}$, and the join and meet operations correspond to set intersection and union, respectively. The abstraction $\sigma_\top$ corresponds to mapping all the symbols in the concrete alphabet to a single symbol, whereas $\sigma_\perp$ corresponds to no abstraction at all. The partial order of $\mathcal{L}_\Sigma$ is defined as follows. Let σ_1, σ_2 be two elements in $\mathcal{L}_\Sigma$,

$$\sigma_1 \sqsubseteq_\Sigma \sigma_2, \text{ if } \sigma_2 \subseteq \sigma_1, \text{ and } \sigma_1 \sqsubset_\Sigma \sigma_2, \text{ if } \sigma_1 \sqsubseteq_\Sigma \sigma_2 \text{ and } \sigma_1 \neq \sigma_2.$$

Let $\sigma_1 \sqsubseteq_\Sigma \sigma_2$. We define the representation function for alphabet abstraction as follows: $\beta_{\sigma_1, \sigma_2} : \Sigma^* \to \Sigma^*$ where

$$\beta_{\sigma_1, \sigma_2}(w) = \{w' \mid |w'| = |w|, \forall i\, 1 \leq i \leq |w|.(w(i) \in \sigma_2 \Rightarrow w'(i) = w(i))$$
$$\wedge (w(i) \notin \sigma_2 \Rightarrow w'(i) = -)\}.$$

The representation function simply maps the symbols that we wish to abstract to the abstract symbol $-$, and maps the rest of the symbols to themselves.

Since the symbolic analysis we defined in Sect. 6.1 uses automata as a symbolic representation, we have to determine how to apply the alphabet abstraction to automata. We define the abstraction function $\alpha_{\sigma_1, \sigma_2}$ on automata using the representation function $\beta_{\sigma_1, \sigma_2}$ as follows: Let A be a single track DFA over σ_1; then $\alpha_{\sigma_1, \sigma_2}(A) = A'$ where A' is a single track DFA over σ_2 such that $\mathcal{L}(A') = \{w \mid \exists w' \in \mathcal{L}(A).\beta_{\sigma_1, \sigma_2}(w') = w\}$.

Note that there may be multiple automata A' that satisfies this constraint. However, since we use minimized multi-track DFAs they will all be equivalent. We define the concretization function $\gamma_{\sigma_2, \sigma_1}$ similarly: Let A be a single track DFA over σ_2; then $\gamma_{\sigma_1, \sigma_2}(A) = A'$ where A' is a single track DFA over σ_1 such that $\mathcal{L}(A') = \{w \mid \exists w' \in \mathcal{L}(A).\beta_{\sigma_1, \sigma_2}(w) = w'\}$.

The definitions we give above are not constructive. We give a constructive definition of the abstraction and concretization functions by first defining an alphabet-abstraction-transducer that maps symbols that we wish to abstract to the abstract symbol $-$, and maps the rest of the symbols to themselves.

An alphabet-abstraction-transducer over σ_1 and σ_2 is a 2-track DFA $A_{\sigma_1, \sigma_2} = \langle Q, \sigma_1 \times \sigma_2, \delta, q_0, F \rangle$, where

- $Q = \{q_0, sink\}, F = \{q_0\}$, and
- $\forall a \in \sigma_2.\delta(q_0, (a, a)) = q_0$,
- $\forall a \in \sigma_1 - \sigma_2.\delta(q_0, (a, -)) = q_0$.

Now, using the alphabet-abstraction-transducer, we can compute the abstraction of a DFA as a post-image computation, and we can compute the concretization of DFA as a pre-image computation. Let A be a single track DFA over σ_1 with track X. $A_{\sigma_1, \sigma_2}(X, X')$ denotes the alphabet transducer over σ_1 and σ_2 where X and X'

correspond to the input and output tracks, respectively. We define the abstraction and concretization functions on automata as (where $X' \mapsto X$ denotes renaming track X' as X):

- $\alpha_{\sigma_1,\sigma_2}(A) \equiv (\exists X.A \sqcap A_{\sigma_1,\sigma_2}(X,X'))[X' \mapsto X]$, and
- $\gamma_{\sigma_1,\sigma_2}(A) \equiv \exists X'.(A[X \mapsto X'] \sqcap A_{\sigma_1,\sigma_2}(X,X'))$.

The definition can be extended to multi-track DFAs. Let A be a multi-track DFA over σ_1^n associated with $\{X_i \mid 1 \leq i \leq n\}$, $\alpha_{\sigma_1^n,\sigma_2^n}(A)$ returns a multi-track DFA over σ_2^n. On the other hand, while A is a multi-track DFA over σ_2^n, $\gamma_{\sigma_1^n,\sigma_2^n}(A)$ returns a multi-track DFA over σ_1^n. We use $A_{\sigma_1^n,\sigma_2^n}$ to denote the extension of the alphabet transducer to multi-track alphabet, where we add $\delta(q_0,(\lambda,\lambda)) = q_0$ to $A_{\sigma_1^n,\sigma_2^n}$ to deal with the padding symbol λ and we use $A_{\sigma_1^n,\sigma_2^n}(X_i,X_i')$ to denote the alphabet transducer associated with tracks X_i and X_i'. The abstraction and concretization of a multi-track DFA A is done track by track as follows:

- $\alpha_{\sigma_1^n,\sigma_2^n}(A) \equiv \forall X_i.(\exists X_i.A \sqcap A_{\sigma_1^n,\sigma_2^n}(X_i,X_i'))[X_i' \mapsto X_i]$, and
- $\gamma_{\sigma_1^n,\sigma_2^n}(A) \equiv \forall X_i.(\exists X_i'.A[X_i \mapsto X_i'] \sqcap A_{\sigma_1^n,\sigma_2^n}(X_i,X_i'))$.

The abstraction lattice $\mathcal{L}_\Sigma$ defines a family of Galois connections between the automata lattices $\mathcal{L}_A$. Each element σ^n in the abstraction lattice $\mathcal{L}_{\Sigma^n}$ is associated with an automata lattice $\mathcal{L}_{\sigma^n}$ corresponding to multi-track automata with the alphabet σ^n. For any pair of elements in the abstraction lattice $\sigma_1^n, \sigma_2^n \in \mathcal{L}_{\Sigma^n}$, if $\sigma_1^n \sqsubseteq_\Sigma \sigma_2^n$, then we can define a Galois connection between the corresponding automata lattices $\mathcal{L}_{\sigma_1^n}$ and $\mathcal{L}_{\sigma_2^n}$ using the abstraction and concretization functions $\alpha_{\sigma_1^n,\sigma_2^n}$ and $\gamma_{\sigma_1^n,\sigma_2^n}$. We formalize this with the following property:
For any Σ^n, and $\sigma_1^n, \sigma_2^n \in \mathcal{L}_{\Sigma^n}$, if $\sigma_1^n \sqsubseteq_\Sigma \sigma_2^n$, the functions $\alpha_{\sigma_1^n,\sigma_2^n}$ and $\gamma_{\sigma_1^n,\sigma_2^n}$ define a Galois connection between the lattices $\mathcal{L}_{\sigma_1^n}$ and $\mathcal{L}_{\sigma_2^n}$ where for any $A_1 \in \mathcal{L}_{\sigma_1^n}$ and $A_2 \in \mathcal{L}_{\sigma_2^n}$:

$$\alpha_{\sigma_1^n,\sigma_2^n}(A_1) \sqsubseteq A_2 \Leftrightarrow A_1 \sqsubseteq \gamma_{\sigma_1^n,\sigma_2^n}(A_2)$$

6.3 Relation Abstraction

In this section, we define the relation abstraction. This abstraction targets the relations between string variables. The abstraction determines the sets of variables that will be analyzed in relation to each other; for each such set the analysis computes a multi-track automaton for each program point such that each track of the automaton corresponds to one variable in that set. In the most abstract case no relations are tracked at all—there is a separate single-track automaton for each variable and the analysis is completely non-relational. On the other hand, in the most precise case we have one single multi-track automaton for each program point.

Let $\overline{X} = \{X_1, \ldots X_n\}$ be a finite set of variables. Let $\chi \subseteq 2^{\overline{X}}$ where $\emptyset \notin \chi$. We say χ defines a relation of $\overline{X}$ if (1) for any $\mathbf{x}, \mathbf{x}' \in \chi$, $\mathbf{x} \not\subseteq \mathbf{x}'$, and (2) $\bigcup_{\mathbf{x} \in \chi} \mathbf{x} = \overline{X}$. The set of χ that defines the relations of $\overline{X}$ form a complete lattice, denoted as $\mathcal{L}_{\overline{X}}$.

- The bottom of the abstraction lattice, denoted as $\chi_\perp$, is $\{\{X_1, X_2, \ldots, X_n\}\}$. This corresponds to the most precise case where, for each program point, a single multi-track automaton is used to represent the set of values for all string variables where each string variable corresponds to one track. This is the representation used in the symbolic reachability analysis described in Chap. 5.
- The top of the abstraction lattice, denoted as $\chi_\top$, is $\{\{X_1\}, \{X_2\}, \{X_3\}, \ldots, \{X_n\}\}$. This corresponds to the most coarse abstraction where, for each program point, n single-track automata are used and each automaton represents the set of values for a single string variable. This corresponds to the approach we used in Chap. 4 [13].

The partial order of the abstraction lattice $\mathcal{L}_{\overline{X}}$ is defined as follows: Let χ_1, χ_2 be two elements in $\mathcal{L}_{\overline{X}}$,

- $\chi_1 \sqsubseteq_{\overline{X}} \chi_2$, if for any $\mathbf{x} \in \chi_2$, there exists $\mathbf{x}' \in \chi_1$ such that $\mathbf{x} \subseteq \mathbf{x}'$.
- $\chi_1 \sqsubset_{\overline{X}} \chi_2$ if $\chi_1 \sqsubseteq_{\overline{X}} \chi_2$ and $\chi_1 \neq \chi_2$.

The automata-based symbolic reachability analysis discussed in earlier chapters can be generalized to a symbolic reachability analysis that works for each abstraction level in the abstraction lattice $\mathcal{L}_{\overline{X}}$. To conduct symbolic reachability analysis for the relation abstraction $\chi \in \mathcal{L}_{\overline{X}}$, we store $|\chi|$ multi-track automata for each program point, where for each $\mathbf{x} \in \chi$, we have a $|\mathbf{x}|$-track DFA, denoted as $A_{\mathbf{x}}$, where each track is associated with a variable in $\mathbf{x}$.

In order to define the abstraction and the concretization functions, we define the following projection and extension operations on automata. For $\mathbf{x}' \subseteq \mathbf{x}$, the projection of $A_{\mathbf{x}}$ to $\mathbf{x}'$, denoted as $A_{\mathbf{x}} \downarrow_{\mathbf{x}'}$, is defined as the $|\mathbf{x}'|$-track DFA that accepts $\{w' \mid w \in \mathcal{L}(A_{\mathbf{x}}), \forall X_i \in \mathbf{x}'.w'[i] = w[i])\}$. Similarly, $\mathbf{x}' \subseteq \mathbf{x}$, the extension of $A_{\mathbf{x}'}$ to $\mathbf{x}$, denoted as $A_{\mathbf{x}'} \uparrow_{\mathbf{x}}$, is defined as the $|\mathbf{x}|$-track DFA that accepts $\{w \mid w' \in \mathcal{L}(A_{\mathbf{x}'}), \forall X_i \in \mathbf{x}'.w[i] = w'[i])\}$.

Let $\mathbf{A}_{\chi} = \{A_{\mathbf{x}} \mid \mathbf{x} \in \chi\}$ be a set of DFAs for the relation χ. The set of string values represented by $\mathbf{A}_{\chi}$ is defined as: $\mathcal{L}(\mathbf{A}_{\chi}) = \mathcal{L}(\bigcap_{\mathbf{x} \in \chi} A_{\mathbf{x}} \uparrow_{\mathbf{x}_u})$, where $\mathbf{x}_u = \{X_1, X_2, \ldots, X_n\}$. I.e., we extend the language of every automaton in $\mathbf{A}_{\chi}$ to all string variables and then take their intersection.

Now, let us define the abstraction and concretization functions for the relation abstraction (which take a set of multi-track automata as input and return a set of multi-track automata as output).

Let $\chi_1 \sqsubseteq_{\overline{X}} \chi_2$; then $\alpha_{\chi_1, \chi_2}(\mathbf{A}_{\chi_1})$ returns a set of DFAs $\{A_{\mathbf{x}'} \mid \mathbf{x}' \in \chi_2\}$, where for each $\mathbf{x}' \in \chi_2$, $A_{\mathbf{x}'} = (\bigcap_{\mathbf{x} \in \chi_1, \mathbf{x}' \cap \mathbf{x} \neq \emptyset} A_{\mathbf{x}} \uparrow_{\mathbf{x}_u}) \downarrow_{\mathbf{x}'}$, where $\mathbf{x}_u = \{X_i \mid X_i \in \mathbf{x}, \mathbf{x} \in \chi_1, \mathbf{x}' \cap \mathbf{x} \neq \emptyset\}$.

$\gamma_{\chi_1, \chi_2}(\mathbf{A}_{\chi_2})$ returns a set of DFAs $\{A_{\mathbf{x}} \mid \mathbf{x} \in \chi_1\}$, where for each $\mathbf{x} \in \chi_1$, $A_{\mathbf{x}} = (\bigcap_{\mathbf{x}' \in \chi_2, \mathbf{x}' \cap \mathbf{x} \neq \emptyset} (A_{\mathbf{x}'} \uparrow_{\mathbf{x}_u})) \downarrow_{\mathbf{x}}$, where $\mathbf{x}_u = \{X_i \mid X_i \in \mathbf{x}', \mathbf{x}' \in \chi_2, \mathbf{x}' \cap \mathbf{x} \neq \emptyset\}$.

Similar to the alphabet abstraction, the relation abstraction lattice $\mathcal{L}_{\overline{X}}$ also defines a family of Galois connections. Each element of the relation abstraction lattice corresponds to a lattice on sets of automata. For each $\chi \in \mathcal{L}_{\overline{X}}$ we define a lattice $\mathcal{L}_{\chi} = (\overline{\mathbf{A}_{\chi}}, \sqsubseteq, \sqcup, \sqcap, \perp, \top)$. Given two sets of automata $\mathbf{A}_{\chi}, \mathbf{A}'_{\chi} \in \overline{\mathbf{A}_{\chi}}$, $\mathbf{A}_{\chi} \sqsubseteq \mathbf{A}'_{\chi}$ if and only if $\mathcal{L}(\mathbf{A}_{\chi}) \subseteq \mathcal{L}(\mathbf{A}'_{\chi})$. The bottom element is defined as $\mathcal{L}(\perp) = \emptyset$

and the top element is defined as $\mathcal{L}(\top) = (\Sigma^n)^*$. The join operator is defined as: $\mathbf{A}_\chi \sqcup \mathbf{A}'_\chi = \{A_x \nabla A'_x \mid \mathbf{x} \in \chi, A_x \in \mathbf{A}_\chi, A'_x \in \mathbf{A}'_\chi\}$ and the meet operator is defined as: $\mathbf{A}_\chi \sqcap \mathbf{A}'_\chi = \{\neg(\neg A_x \nabla \neg A'_x) \mid \mathbf{x} \in \chi, A_x \in \mathbf{A}_\chi, A'_x \in \mathbf{A}'_\chi\}$.

For any pair of elements in the relation abstraction lattice $\chi_1, \chi_2 \in \mathcal{L}_{\overline{X}}$, if $\chi_1 \sqsubseteq_{\overline{X}}$ χ_2, then the abstraction and concretization functions α_{χ_1,χ_2} and γ_{χ_1,χ_2} define a Galois connection between $\mathcal{L}_{\chi_1}$ and $\mathcal{L}_{\chi_2}$. We formalize this with the following property: For any $\chi_1, \chi_2 \in \mathcal{L}_{\overline{X}}$, if $\chi_1 \sqsubseteq_{\overline{X}} \chi_2$, then the functions α_{χ_1,χ_2} and γ_{χ_1,χ_2} define a Galois connection between $\mathcal{L}_{\chi_1}$ and $\mathcal{L}_{\chi_2}$ where for any $\mathbf{A}_{\chi_1} \in \mathcal{L}_{\chi_1}$ and $\mathbf{A}'_{\chi_2} \in \mathcal{L}_{\chi_2}$:

$$\alpha_{\chi_1,\chi_2}(\mathbf{A}_{\chi_1}) \sqsubseteq \mathbf{A}'_{\chi_2} \Leftrightarrow \mathbf{A}_{\chi_1} \sqsubseteq \gamma_{\sigma_1^n,\sigma_2^n}(\mathbf{A}'_{\chi_2})$$

6.4 Composing Abstractions

As shown in the two previous sections, both alphabet and relation abstractions form abstraction lattices which allow different levels of abstraction. Combining these abstractions leads to a product lattice where each point in this lattice corresponds to the combination of a particular alphabet abstraction with a particular relation abstraction. This creates an even larger set of Galois connections, one for each possible combination of alphabet and relation abstractions. Given Σ and $\overline{X} = \{X_1, \ldots, X_n\}$, we define a point in this product lattice as an *abstraction class* which is a pair (χ, σ) where $\chi \in \mathcal{L}_{\overline{X}}$ and $\sigma \in \mathcal{L}_{\Sigma}$. The abstraction classes of $\overline{X}$ and Σ also form a complete lattice, of which the partial order is defined as: $(\chi_1, \sigma_1) \sqsubseteq (\chi_2, \sigma_2)$ if $\chi_1 \sqsubseteq \chi_2$ and $\sigma_1 \sqsubseteq \sigma_2$.

Given Σ and $\overline{X} = \{X_1, \ldots, X_n\}$, we can select any abstraction class in the product lattice during our analysis. The selected abstraction class (χ, σ) determines the precision and efficiency of our analysis. If we select the abstraction class $(\chi_\perp, \sigma_\perp)$, we conduct our most precise relational string analysis. The relations among $\overline{X}$ will be kept using one n-track DFA at each program point. If we select $(\chi_\top, \sigma_\top)$, we only keep track of the *length* of each string variable individually. Although we abstract away almost all string relations and contents in this case, this kind of path-sensitive (w.r.t length conditions on a single variable) size analysis can be used to detect buffer overflow vulnerabilities [36, 115]. If we select $(\chi_\perp, \sigma_\top)$, then we will be conducting relational size analysis. Finally, earlier string analysis techniques that use DFAs, such as [13, 130], correspond to the abstraction class $(\chi_\top, \sigma_\perp)$, where multiple single-track DFAs over Σ are used to encode reachable states. As shown in [13, 126, 130], this type of analysis is useful for detecting XSS and SQL Injection vulnerabilities.

Figure 6.9 summarizes the different types of abstractions that can be obtained using the abstraction framework we defined. The alphabet and relation abstractions can be seen as two knobs that determine the level of the precision of the string analysis. The alphabet abstraction knob determines how much of the string content

Fig. 6.9 Some abstractions
from the abstraction lattice
and corresponding analyses

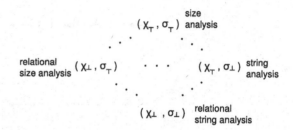

is abstracted away. In the limit, the only information left about the string values
is their lengths. On the other hand, the relation abstraction knob determines which
set of variables should be analyzed in relation to each other. In the limit, all values
are projected to individual variables. Different abstraction classes can be useful in
different cases.

6.5 Automata Widening Operation

The symbolic reachability analysis is a least fixpoint computation. The process
terminates when a fixpoint is reached. Since string systems are infinite state systems,
the iterative reachability computation is not guaranteed to terminate. We incorporate
an automata widening operator that was proposed in [17], to accelerate the fixpoint
computation. During the widening operation, we partition the states of the automata
to equivalence classes according to specific equivalence conditions and merge states
in the same equivalence class. We use the same equivalence conditions defined
in [17] in our implementation of the widening operator.

Given two finite automata $A = \langle Q, q_0, \Sigma, \delta, F \rangle$ and $A' = \langle Q', q_0', \Sigma, \delta', F' \rangle$, we
first define the binary relation $\equiv_\nabla$ on $Q \cup Q'$ as follows. Given $q \in Q$ and $q' \in Q'$,
we say that $q \equiv_\nabla q'$ and $q' \equiv_\nabla q$ if and only if

$$\forall w \in \Sigma^*. \, \delta^*(q, w) \in F \Leftrightarrow \delta'^*(q', w) \in F'. \tag{6.1}$$

$$\text{or} \quad q, q' \neq sink \wedge \exists w \in \Sigma^*. \, \delta^*(q_0, w) = q \wedge \delta'^*(q_0', w) = q', \tag{6.2}$$

where $\delta^*(q, w)$ is defined as the state that A reaches after consuming w starting from
state q. In other words, condition 6.1 states that $q \equiv_\nabla q'$ if $\forall w \in \Sigma^*$, w is accepted
by A from q then w is accepted by A' from q', and vice versa. Condition 6.2 states
that $q \equiv_\nabla q'$ if $\exists w \in \sigma$, A reaches state q and A' reaches state q' after consuming w
from its initial state. For $q_1, q_2 \in Q$ and $q_1 \neq q_2$ we say that $q_1 \equiv_\nabla q_2$ if and only if

$$\exists q' \in Q'. \, q_1 \equiv_\nabla q' \wedge q_2 \equiv_\nabla q' \, \vee \, \exists q \neq q_1, q_2 \in Q. \, q_1 \equiv_\nabla q \wedge q_2 \equiv_\nabla q \tag{6.3}$$

Similarly we can define $q_1' \equiv_\nabla q_2'$ for $q_1' \in Q'$ and $q_2' \in Q'$.

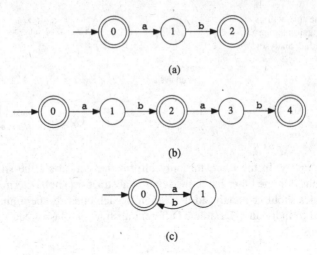

Fig. 6.10 Widening automata. (a) A. (b) A'. (c) $A \nabla A'$

It can be seen that $\equiv_\nabla$ is an equivalence relation. Let C be the set of equivalence classes of $\equiv_\nabla$. We define $A \nabla A' = \langle Q'', q_0'', \Sigma, \delta'', F'' \rangle$ by:

$$Q'' = C$$

$$q_0'' = c \text{ s.t. } q_0 \in c \wedge q_0' \in c$$

$$\delta''(c_i, \sigma) = c_j \text{ s.t.}$$

$$(\forall q \in c_i \cap Q. \; \delta(q, \sigma) \in c_j \vee \delta(q, \sigma) = sink) \wedge$$

$$(\forall q' \in c_i \cap Q'. \; \delta'(q', \sigma) \in c_j \vee \delta'(q', \sigma) = sink)$$

$$F'' = \{ c \mid \exists q \in F \cup F'. \; q \in c \}$$

In other words, the set of states of $A \nabla A'$ is the set C of equivalence classes of $\equiv_\nabla$. Transitions are defined from the transitions of A and A'. The initial state is the class containing the initial states q_0 and q_0'. The set of final states is the set of classes that contain some of the final states in F and F'. It can be shown that, given two automata A and A', $\mathcal{L}(A) \cup \mathcal{L}(A') \subseteq \mathcal{L}(A \nabla A')$ [17].

In Fig. 6.10, we give an example for the widening operation. $\mathcal{L}(A) = \{\epsilon, ab\}$ and $\mathcal{L}(A') = \{\epsilon, ab, abab\}$. The set of equivalence classes is $C = \{q_0'', q_1''\}$, where $q_0'' = \{q_0, q_0', q_2, q_2', q_4'\}$ and $q_1'' = \{q_1, q_1', q_3'\}$. $\mathcal{L}(A \nabla A') = (ab)^*$.

As we discussed in earlier chapters, we use this widening operator iteratively to compute an over-approximation of the least fixpoint that corresponds to the reachable values of string expressions. To simplify the discussion, let us assume a program with a single string variable represented with one automaton A. Let A_i represent the automaton computed at the ith iteration and let I denote the set of initial values of the string variable. The fixpoint computation will compute a sequence A_0, $A_1, \ldots, A_i, \ldots$, where $\mathcal{L}(A_0) = I$ and $\mathcal{L}(A_i) = \mathcal{L}(A_{i-1}) \cup \mathcal{L}(\text{POST}(A_{i-1}))$ where the post-image for different statements can be computed as described in Chap. 4.

Formally speaking, $\mathcal{L}(A)$ is a fixpoint if $\mathcal{L}(A) = \mathcal{L}(A) \cup \mathcal{L}(\text{POST}(A))$. $\mathcal{L}(A_\infty)$ is the least fixpoint if $\mathcal{L}(A_\infty)$ is a fixpoint and for all other fixpoint $\mathcal{L}(A)$, $\mathcal{L}(A_\infty) \subseteq \mathcal{L}(A)$.

We reach the least fixpoint $\mathcal{L}(A_\infty) = \mathcal{L}(A_i)$ at the ith iteration when $\mathcal{L}(A_i) = \mathcal{L}(A_{i-1})$. Since we are dealing with an infinite state system, the fixpoint computation may not converge.

Given the widening operator, we actually compute a sequence A'_0, A'_1, …, A'_i, …, that over-approximates the fixpoint computation where A'_i is defined as: $\mathcal{L}(A'_0) = \mathcal{L}(A_0)$, and for $i > 0$, $\mathcal{L}(A'_i) = \mathcal{L}(A'_{i-1} \nabla A_i)$, where $\mathcal{L}(A_i) = \mathcal{L}(A'_{i-1}) \cup \mathcal{L}(\text{POST}(A'_{i-1}))$.

Let A'_∞ denote the limit of this approximate sequence where there exists a j, $\mathcal{L}(A'_\infty) = \mathcal{L}(A'_j) = \mathcal{L}(A'_{j-1})$. Then we have the following result from [17]. We say $A_1 = \langle Q_1, q_{10}, \Sigma, \delta_1, F_1 \rangle$ is simulated by $A_2 = \langle Q_2, q_{20}, \Sigma, \delta_2, F_2 \rangle$ if and only if there exists a total function $f : Q_1 \setminus \{sink\} \to Q_2$ such that $\delta_1(q, \sigma) = sink$ or $f(\delta_1(q, \sigma)) = \delta_2(f(q), \sigma)$ for all $q \in Q_1 \setminus \{sink\}$ and $\sigma \in \Sigma$. Furthermore, $f(q_{10}) = q_{20}$ and for all $q \in F_1, f(q) \in F_2$. We say $A = \langle Q, q_0, \Sigma, \delta, F \rangle$ is state-disjoint if and only if for all $q \neq q' \in Q$, $\mathcal{L}(q) \neq \mathcal{L}(q')$, where $\mathcal{L}(k) = \{w \mid \delta(k, w) \in F\}$, for $k \in Q$.

It has been shown that if (1) A_∞ exists, (2) A_∞ is a state-disjoint automaton, and (3) A_0 is simulated by A_∞, then if A'_∞ exists (i.e., if the approximate sequence converges) then $\mathcal{L}(A'_\infty) = \mathcal{L}(A_\infty)$.

Consider a simple program where a variable initialized to the empty string and in each iteration of a loop the substring ab is concatenated to the variable. When we use the symbolic forward reachability algorithm, the exact sequence $A_0, A_1, …,$ $A_i, …$ will never converge to the least fixpoint, where $\mathcal{L}(A_0) = \{\epsilon\}$ and $\mathcal{L}(A_i) = \{(ab)^k \mid 0 \leq k \leq i\}$. However, A_∞ exists and $\mathcal{L}(A_\infty) = (ab)^*$. In addition, A_∞ is a state-disjoint automaton, and A_0 is simulated by A_∞. These conditions imply that once the computation of the approximate sequence reaches the fixpoint, the fixpoint is equal to A_∞ and the analysis is precise. Computation of the approximate sequence is shown in Fig. 6.11. $\mathcal{L}(A'_i) = \mathcal{L}(A'_{i-1} \nabla A_i)$, where $\mathcal{L}(A_i) = \mathcal{L}(A'_{i-1}) \cup \mathcal{L}(\text{POST}(A'_{i-1}))$ and $\text{POST}(A)$ returns an automaton that accepts $\{wab \mid w \in \mathcal{L}(A)\}$. In this case, we reach the fixpoint at the second iteration and $A'_\infty = A_\infty = A'_2$.

A more general case that we commonly encounter in real programs is that we start from a set of initial strings (accepted by A_{init}), and concatenate an arbitrary but fixed set of strings (accepted by A_{tail}) at each iteration. Based on the discussion above, one can conclude that if the DFA A that accepts $\mathcal{L}(A_{init})\mathcal{L}(A_{tail})^*$ is state-disjoint, then our analysis via widening will reach the precise least fixpoint when it terminates.

Fig. 6.11 The approximate sequence of a converging example. (a) A'_0. (b) A_1. (c) A'_1 and A'_2

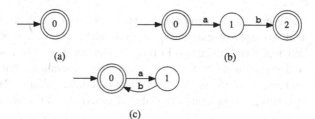

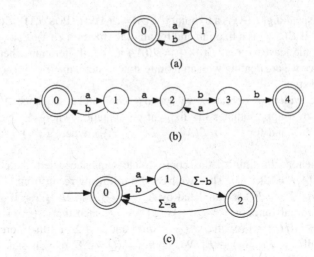

Fig. 6.12 The approximate sequence of a non-regular example. (a) A_1'. (b) A_2'. (c) A_2''

The automata widening operator defined in [17] has two variations and only the coarser version guarantees convergence. The coarser widening operator ∇_c is defined the same as ∇ except that we discard the condition $q, q' \neq sink$ in Eq. (6.2). In our implementation we start with the more precise version (∇) and after a constant number of steps switch to the coarser version (∇_c) to guarantee convergence.

In general, the least fixpoint $\mathcal{L}(A_\infty)$ may not exist, or even if it exists, the language may not be regular. For instance, if we change the previous example to that POST(A) returns an automaton that accepts $\{awb \mid w \in \mathcal{L}(A)\}$ instead, we will have $\mathcal{L}(A_i) = \{a^k b^k \mid 0 \leq k \leq i\}$. Though the least fixpoint $\mathcal{L}(A_\infty) = \{a^n b^n \mid 0 \leq n\}$ exists, it is not a regular language. Let $\mathcal{L}(A_0) = \mathcal{L}(A_0') = \{\epsilon\}$. For the above example, $A_1' = A_0' \nabla (A_0' \sqcup \text{POST}(A_0'))$ and $A_2' = A_1' \nabla (A_1' \sqcup \text{POST}(A_1'))$ are shown in Fig. 6.12a, b, respectively. The sequence does not converge. On the other hand, if we apply the coarser widening operator at the second iteration, we get $A_2'' = A_1' \nabla_c (A_1' \sqcup \text{POST}(A_1'))$ shown in Fig. 6.12c, and reach the fixpoint at the next iteration. The result is an over approximation of the least fixpoint $\mathcal{L}(A_\infty)$.

6.6 Summary

Verifying string manipulating programs is an undecidable problem in general and any approximate string analysis technique has an inherent tension between efficiency and precision. In this Chapter we discussed a set of sound abstractions and approximations for strings and string operations that allow for both efficient and precise verification of string manipulating programs. Particularly, we introduced two string abstractions—alphabet abstraction and relation abstraction—that can be used in combination to tune the analysis precision and efficiency. We showed that these abstractions form an abstraction lattice that generalizes the string analysis techniques studied previously in isolation, such as size analysis or non-relational string analysis. We also discussed the widening operation to approximate the set of states that are characterized by automata. Widening is crucial for accelerating fixpoint computations and achieving convergence.

Chapter 7
Constraint-Based String Analysis

Analysis of string manipulating programs has been studied extensively in recent years. One of the commonly used program analysis technique, *symbolic execution*, can be applied to string manipulating programs. However, symbolic execution of string manipulating programs is difficult since solving string constraints is a challenging problem. String constraint solving is challenging due to two main reasons: (1) With the increasing usage of strings in modern software development, programming languages provide increasingly complex string operations that need to be handled by string constraint solvers. (2) String constraints are usually mixed with integer constrains which requires solving integer constraints together with string constraints. In this chapter, we first provide an illustrative example of the symbolic execution of a string manipulating program. Then, we discuss the details of automata-based constraint solving and model counting extension to string constraint solving.

7.1 Symbolic Execution with String Constraints

Symbolic execution is a program analysis technique to determine what input values cause each path of a program to execute [65]. Symbolic execution assumes symbolic values for the program inputs rather than using concrete values as normal execution of the program would. Expressions encountered during symbolic execution are expressed as functions of the symbolic variables. At any point during symbolic execution, program state is described with the value of the program counter and with a symbolic expressions known as the path condition (PC). A PC is a constraint on input values that must be satisfied in order for a program to reach the location that PC corresponds. The set of all possible executions of a program is represented by a symbolic execution tree.

© Springer International Publishing AG 2017

T. Bultan et al., *String Analysis for Software Verification and Security*,
https://doi.org/10.1007/978-3-319-68670-7_7

Fig. 7.1 A Java string
manipulation example

```
 1  public void site_exec(String cmd) {
 2      String p = "home/ftp/bin";
 3      int j, sp = cmd.indexOf(' ');
 4
 5      if (sp == -1) {
 6          j = cmd.lastIndexOf('/');
 7      } else {
 8          j = cmd.lastIndexOf('/', sp);
 9      }
10
11      String r = cmd.substring(j);
12      int l = r.length() + p.length();
13
14      if (l > 32) {
15          return;
16      }
17
18      String buf = p + r;
19      boolean t = buf.contains("%n");
20
21      if (t == true) {
22          throw new Exception("THREAT");
23      }
24
25      execute(buf);
26      return;
27  }
```

Figure 7.1 shows a JAVA string manipulation example originally presented
as a command injection example in an earlier work [89] and used as a string
constraint solving example in Symbolic Path Finder[1] (SPF). It is part of the WU-
FTPD implementation of the file transfer protocol. It is originally written in C
programming language and the example is converted from the original version. If
the input command contains substring "%n" an exception is thrown at line 22. In the
original implementation, this situation would allow user to alter program stack and
gain privileged access to server running the program. This example demonstrates
one application of symbolic execution, it checks feasibility of program execution
paths that may lead to a vulnerability.

Figure 7.2 represents the symbolic execution tree of site_exec function
in the example. Symbolic execution tree is built on the fly based on a traversal
strategy (e.g., depth-first, breadth-first). Symbolic execution tree of our running
example is created by exploring program paths with depth-first exploration strategy.

---

[1]http://babelfish.arc.nasa.gov/trac/jpf/wiki/projects/jpf-symbc.

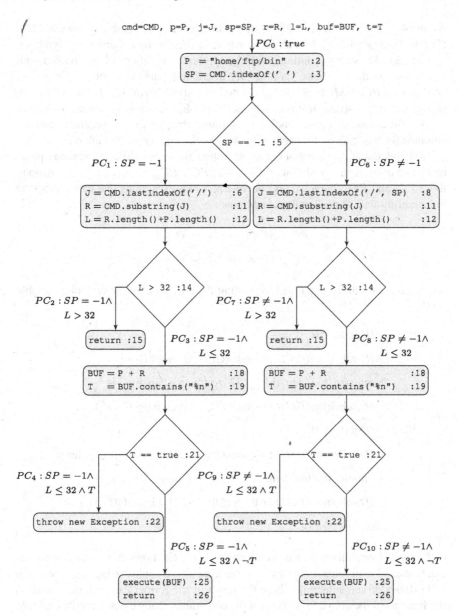

Fig. 7.2 Symbolic execution tree of the example code in Fig. 7.1

Rectangles represent updates on string expressions and diamonds represent branch points encountered during symbolic execution. Initially, all program variables are replaced with symbolic variables, cmd=CMD, p=P, j=J, sp=SP, r=R, l=L, buf=BUF, t=T and initial PC is set to true, $PC_0 = true$. At line 3 symbolic variable P is assigned to concrete string with value "home/ftp/bin".

At line 4 symbolic variable SP is assigned to a function of symbolic variable CMD. Line 6 corresponds to a branch point where it checks the condition on symbolic variable SP. In order to continue with the symbolic execution, PC is updated with constraints on the symbolic variables for alternative paths, i.e., $PC_1 : SP = -1$ and later on $PC_6 : SP \neq -1$ are generated. Satisfiability of PC_1 is checked using string constraint solvers; if it is satisfiable, symbolic execution continues to explore deeper. Otherwise, if a path condition is unsatisfiable, symbolic execution does not continue for that path and it backtracks and checks satisfiability of alternative PCs.

In the example, the symbolic execution tree shows six different feasible paths represented by path constraints PC_2, PC_4, PC_5, PC_7, PC_9, and PC_{10}. Among those PC_4 and PC_9 characterizes concrete program executions that can exploit the vulnerability with the following conditions:

$$PC_4 : SP = -1 \wedge L \leq 32 \wedge T$$
$$PC_9 : SP \neq -1 \wedge L \leq 32 \wedge T$$

where each PC can be expanded by writing them as function of symbolic variable CMD:

$PC_4 :$ CMD.indexOf$('\ ') = -1 \wedge$

CMD.substring$(CMD$.lastIndexOf$('/'))$.length$() +$

"home/ftp/bin".length$() \leq 32 \wedge$ ("home/ftp/bin"$+$

CMD.substring$(CMD$.lastIndexOf$('/')))$.contains("%n")

$PC_9 :$ CMD.indexOf$('\ ') \neq -1 \wedge$

CMD.substring$(CMD$.lastIndexOf$('/', CMD$.indexOf$('\ ')))$.length$()$

$+$ "home/ftp/bin".length$() \leq 32 \wedge$ ("home/ftp/bin"$+$

CMD.substring$(CMD$.lastIndexOf$('/', CMD$.indexOf$('\ '))))$

.contains("%n")

Expanded versions of path constraints PC_4 and PC_9 shows that string constraints can be mixed with integer constraints. There are also complex string functions such as lastIndexOf and substring where the result of the former can be a parameter to the latter as in the above PCs. Type of the constraints that we get from the example shows that string constraint solving is essential for symbolic execution of string manipulating programs.

7.2 Automata-Based String Constraint Solving

In the previous chapters we discussed how to construct automata for concatenation and replace functions and word equations. In this section we discuss how to construct automata for string constraints in general. Given a constraint and a variable, our goal is to construct an automaton that accepts all strings, which, when assigned as the value of the variable in the given constraint, results in a satisfiable constraint. This construction allows us to compute the post condition of both:

- *Branch statements:* Since branch conditions are boolean expressions, i.e., constraints, and the post condition of a branch condition is the set of values that satisfy the corresponding constraint; and
- *Assignment statements:* Since we can rewrite the assignment statement $v :=$ *sexp*; as an equality $v' = sexp$ where v' denotes the value of the variable v after the assignment statement is executed, and computing the set of values for v' that satisfies this constraint is equivalent the set of values v can take in the post condition of the assignment statement.

Given a constraint F, let V_F denote the set of variables that appear in F. Let $F[s/v]$ denote the constraint that is obtained from F by replacing all appearances of $v \in V_F$ with the string constant s. We define the truth set of the formula F for variable v as $[\![F, v]\!] = \{s \mid F[s/v]$ is satisfiable$\}$.

We identify three classes of constraints:

1. *Single-variable constraints* are constructed using at most one string variable (i.e., $V_F = \{v\}$ or $V_F = \emptyset$).
2. *Pseudo-relational constraints* are a set of constraints that we define in the next section, for which the truth sets are regular (i.e., each $[\![F, v]\!]$ is a regular set).
3. *Relational constraints* are the constraints that are not pseudo-relational constraints (truth sets of relational constraints can be non-regular).

7.2.1 Mapping Constraints to Automata

Given a constraint F and a variable v, our goal is to construct an automaton A, such that $\mathcal{L}(A) = [\![F, v]\!]$. In this section, we focus on mapping string constraints to single-track automata. In the next section we discuss constructing multi-track automata for relational constraints.

7.2.1.1 Automata Construction for Single-Variable Constraints

Let us define an automata constructor function $\mathcal{A}$ such that, given a formula F and a variable v, $\mathcal{A}(F, v)$ is an automaton where $\mathcal{L}(\mathcal{A}(F, v)) = [\![F, v]\!]$. In this section we discuss how to implement the automata constructor function $\mathcal{A}$.

Fig. 7.3 Syntax tree of the
formula
$F \equiv \neg \mathrm{match}(x, (01)^*) \wedge$
$\mathrm{length}(x) \geq 1$

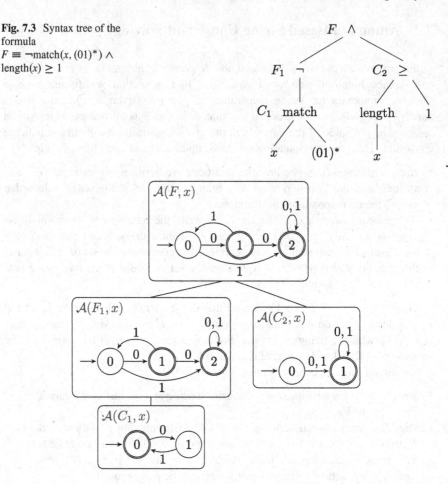

Fig. 7.4 The automata construction that traverses the syntax tree from the leaves towards the root

Consider the following string constraint $F \equiv \neg(x \in (01)^*) \wedge \mathrm{length}(x) \geq 1$ over
the alphabet $\Sigma = \{0, 1\}$. Let us name the sub-constraints of F as $C_1 \equiv x \in (01)^*$,
$C_2 \equiv \mathrm{length}(x) \geq 1$, $F_1 \equiv \neg C_1$, where $F \equiv F_1 \wedge C_2$. The automata construction
algorithm starts from the basic constraints at the leaves of the syntax tree (C_1 and C_2
in Fig. 7.3), and constructs the automata for them. Then it traverses the syntax tree
towards the root by constructing an automaton for each node using the automata
constructed for its children (where the automaton for F_1 is constructed using the
automaton for C_1 and the automaton for F is constructed using the automata for
F_1 and C_2). Figure 7.4 demonstrates the automata construction algorithm on our
running example.

Let $\mathcal{A}(\Sigma^*)$, $\mathcal{A}(\Sigma^n)$, $\mathcal{A}(s)$, and $\mathcal{A}(\emptyset)$ denote automata that accept the languages
Σ^*, Σ^n, $\{s\}$, and $\emptyset$, respectively. We construct the automaton $\mathcal{A}(F, v)$ recursively
on the structure of the single-variable constraint F as follows:

- case $V_F = \emptyset$ (i.e., there are no variables in F): Evaluate the constraint F. If $F \equiv$ **true** then $\mathcal{A}(F, v) = \mathcal{A}(\Sigma^*)$, otherwise $\mathcal{A}(F, v) = \mathcal{A}(\emptyset)$.
- case $F \equiv \neg F_1$: $\mathcal{A}(F, v)$ is constructed using $\mathcal{A}(F_1, v)$ and it is an automaton that accepts the complement language $\Sigma^* - \mathcal{L}(\mathcal{A}(F_1, v))$.
- case $F \equiv F_1 \wedge F_2$ or $F \equiv F_1 \vee F_2$: $\mathcal{A}(F, v)$ is constructed using $\mathcal{A}(F_1, v)$ and $\mathcal{A}(F_2, v)$ using automata product, and it accepts the language $\mathcal{A}(F_1, v) \sqcap \mathcal{A}(F_2, v)$ or $\mathcal{A}(F_1, v) \sqcup \mathcal{A}(F_2, v)$, respectively.
- case $F \equiv$ match(v, R): $\mathcal{A}(F, v)$ is constructed using regular expression to automata conversion algorithm and accepts all strings that match the regular expression R.
- case $F \equiv v = s$: $\mathcal{A}(F, v) = \mathcal{A}(s)$.
- case $F \equiv$ length$(v) = n$: $\mathcal{A}(F, v) = \mathcal{A}(\Sigma^n)$.
- case $F \equiv$ length$(v) < n$: $\mathcal{A}(F, v)$ is an automaton that accepts the language $\{\varepsilon\} \cup \Sigma^1 \cup \Sigma^2 \cup \ldots \cup \Sigma^{n-1}$.
- case $F \equiv$ length$(v) > n$: $\mathcal{A}(F, v)$ is constructed using $\mathcal{A}(\Sigma^{n+1})$ and $\mathcal{A}(\Sigma^*)$ and then constructing an automaton that accepts the concatenation of those languages, i.e., $\Sigma^{n+1}\Sigma^*$.
- case $F \equiv$ contains(v, s): $\mathcal{A}(F, v)$ is an automaton that is constructed using $\mathcal{A}(\Sigma^*)$ and $\mathcal{A}(s)$ and it accepts the language $\Sigma^* s \Sigma^*$.
- case $F \equiv$ begins(v, s): $\mathcal{A}(F, v)$ is constructed using $\mathcal{A}(\Sigma^*)$ and $\mathcal{A}(s)$, and it accepts the language $s\Sigma^*$.
- case $F \equiv$ ends(v, s): $\mathcal{A}(F, v)$ is constructed using $\mathcal{A}(\Sigma^*)$ and $\mathcal{A}(s)$, and it accepts the language $\Sigma^* s$.
- case $F \equiv n =$ indexof(v, s): Let L_i denote the language $\Sigma^i s \Sigma^*$. Automata that accept the languages L_i can be constructed using $\mathcal{A}(\Sigma^i)$, $\mathcal{A}(s)$, and $\mathcal{A}(\Sigma^*)$. Then $\mathcal{A}(F, v)$ is the automaton that accepts the language $\Sigma^n s \Sigma^* - (\{\varepsilon\} \cup L_1 \cup L_2 \cup \ldots \cup L_{n-1})$ which can be constructed using $\mathcal{A}(\Sigma^n)$, $\mathcal{A}(s)$, $\mathcal{A}(\Sigma^*)$, and the automaton that accepts L_i.

7.2.1.2 Pseudo-Relational Constraints

Pseudo-relational constraints are multi-variable constraints. Note that, using multiple variables, one can specify constraints with non-regular truth sets. For example, given the constraint $F \equiv x = y \,.\, y$, $[\![F, x]\!]$ is not a regular set, so we cannot construct an automaton precisely recognizing its truth set. Below, we define a class of constraints called pseudo-relational constraints for which $[\![F, v]\!]$ is regular.

We assume that constraint F is converted to DNF form where $F \equiv \vee_{i=1}^{n} F_i$, $F_i \equiv \wedge_{j=1}^{m} C_{ij}$, and each C_{ij} is either a basic constraint or negation of a basic constraint. The constraint F is pseudo-relational if each F_i is pseudo-relational.

Given $F \equiv C_1 \wedge C_2 \wedge \ldots \wedge C_n$, where each C_i is either a basic constraint or negation of a basic constraint, for each C_i, let V_{C_i} denote the set of variables that appear in C_i. We call F pseudo-relational if the following conditions hold:

1. Each variable $v \in V_F$ appears in each C_i at most once.
2. There is only one variable, $v \in V_F$, that appears in more than one constraint C_i where $v \in V_{C_i} \wedge |V_{C_i}| > 1$, and in each C_i that v appears in, v is on the left hand side of the constraint. We call v the *projection variable*.
3. For all variables $v' \in V_F$ other than the projection variable, there is a single constraint C_i where $v' \in V_{C_i} \wedge |V_{C_i}| > 1$ and the projection variable v appears in C_i, i.e., $v \in V_{C_i}$.
4. For all constraints C_i where $|V_{C_i}| > 1$, C_i is not negated in the formula F.

Many string constraints extracted from programs via symbolic execution are pseudo-relational constraints, or can be converted to pseudo-relational constraints. The projection variable represents either the variable that holds the value of the user's input to the program (for example, user input to a web application that needs to be validated), or the value of the string expression at a program sink. A program sink is a program point (such as a security sensitive function) for which it is necessary to compute the set of values that reach to that program point in order to check for vulnerabilities.

For example, following constraint is a pseudo-relational constraint extracted from a web application (regular expressions are simplified):

$$(x = y \cdot z) \wedge (\text{length}(y) = 0) \wedge \neg(z \in (0|1)) \wedge (x = t) \wedge \neg(t \in 0^*)$$

7.2.1.3 Automata Construction for Pseudo-Relational Constraints

Given a pseudo-relational constraint F and the projection variable v, we now discuss how to construct the automaton $\mathcal{A}(F, v)$ that accepts $[\![F, v]\!]$. As above, we assume that F is converted to DNF form where $F \equiv \vee_{i=1}^{n} F_i$, $F_i \equiv \wedge_{j=1}^{m} C_{ij}$, and each C_{ij} is either a basic constraint or negation of a basic constraint.

In order to construct the automaton $\mathcal{A}(F, v)$ we first construct the automata $\mathcal{A}(F_i, v)$ for each F_i where $\mathcal{A}(F_i, v)$ accepts the language $[\![F_i, v]\!]$. Then we combine the $\mathcal{A}(F_i, v)$ automata using automata product such that $\mathcal{A}(F, v)$ accepts the language $[\![F_1, v]\!] \cup [\![F_2, v]\!] \cup \ldots \cup [\![F_m, v]\!]$.

Since we discussed how to handle disjunction, from now on we focus on constraints of the form $F \equiv C_1 \wedge C_2 \wedge \ldots \wedge C_n$ where each C_i is either a basic constraint or negation of a basic constraint. For each C_i, let V_{C_i} denote the set of variables that appear in C_i. If V_{C_i} is a singleton set, then we refer to the variable in it as v_{C_i}.

First, for each single-variable constraint C_i that is not negated, we construct an automaton that accepts the truth set of the constraint C_i, $[\![C_i, v_{C_i}]\!]$, using the techniques we discussed above for single-variable constraints. If C_i is negated, then we construct the automaton that accepts the complement language $\Sigma^* - [\![C_i, v_{C_i}]\!]$ (note that, only single-variable constraints can be negated in pseudo-relational constraints). Let us call these automata $\mathcal{A}(C_i, v_{C_i})$ (some of which may correspond to negated constraints).

Then, for any variable $v' \in V_F$ that is not the projection variable, we construct an automaton $\mathcal{A}(F, v')$ which accepts the intersection of the languages $\mathcal{A}(C_i, v')$ for all single-variable constraints that v' appears in, i.e.,

$$\mathcal{L}(\mathcal{A}(F, v')) = \bigcap_{V_{C_i} = \{v'\}} \mathcal{L}(\mathcal{A}(C_i, v')).$$

Next, for each multi-variable constraint C_i we construct an automaton that accepts the language $[\![C_i, v]\!]$ where v is the projection variable as follows:

- case $C_i \equiv v = v'$: $\mathcal{A}(C_i, v) = \mathcal{A}(F, v')$.
- case $C_i \equiv v = v_1 . v_2$: $\mathcal{A}(C_i, v)$ is constructed using the automata $\mathcal{A}(F, v_1)$ and $\mathcal{A}(F, v_2)$ and it accepts the concatenation of the languages $\mathcal{L}(\mathcal{A}(F, v_1))$ and $\mathcal{L}(\mathcal{A}(F, v_2))$.
- case $C_i \equiv \text{length}(v) = \text{length}(v')$: Given the automaton $\mathcal{A}(F, v')$, we construct an automaton $A_{\text{length}(F,v')}$ such that $s \in \mathcal{L}(A_{\text{length}(F,v')}) \Leftrightarrow \exists s' : \text{length}(s) = \text{length}(s') \wedge s' \in \mathcal{L}(\mathcal{A}(F, v'))$. Then, $\mathcal{A}(C_i, v) = A_{\text{length}(F,v')}$.
- case $C_i \equiv \text{length}(v) < \text{length}(v')$: Given the automaton $\mathcal{A}(F, v')$ we find the length of the maximum word accepted by $\mathcal{A}(F, v')$, which is infinite if $\mathcal{A}(F, v')$ has a loop that can reach an accepting state. If it is infinite then $\mathcal{A}(C_i, v) = A(\Sigma^*)$. If not, then given the maximum length m, $\mathcal{A}(C_i, v)$ is the automaton that accepts the language $\{\varepsilon\} \cup \Sigma^1 \cup \Sigma^2 \cup \ldots \cup \Sigma^{m-1}$. Note that if $m = 0$ then $\mathcal{A}(C_i, v) = A(\emptyset)$.
- case $C_i \equiv \text{length}(v) > \text{length}(v')$: Given the automaton $\mathcal{A}(F, v')$ we find the length of the minimum word accepted by $\mathcal{A}(F, v')$. Given the minimum length m, $\mathcal{A}(C_i, v)$ is the automaton that accepts the concatenation of the languages accepted by $A(\Sigma^{m+1})$ and $A(\Sigma^*)$, i.e, $\Sigma^{m+1} \Sigma^*$.
- case $C_i \equiv v = \text{replace}(v', s, s)$: Given the automaton $\mathcal{A}(F, v')$ we use the construction presented in [129, 130] for language based replacement to construct the automaton $\mathcal{A}(C_i, v)$.

The final step of the construction is to construct $\mathcal{A}(F, v)$ using the automata $\mathcal{A}(C_i, v)$ where $\mathcal{L}(\mathcal{A}(F, v)) = \bigcap_{v \in V_{C_i}} \mathcal{L}(\mathcal{A}(C_i, v))$.

For pseudo-relational constraints, the automaton $\mathcal{A}(F, v))$ constructed based on the above construction accepts the truth set of the formula F for the projected variable, i.e., $\mathcal{L}(\mathcal{A}(F, v)) = [\![F, v]\!]$. However, the replace function has different variations in different programming languages (such as first-match versus longest-match replace) and the match pattern can be given as a regular expression. The language-based replace automata construction we use [129, 130] over-approximates the replace operation in some cases, which would then result in over-approximation of the truth set: $\mathcal{L}(\mathcal{A}(F, v)) \supseteq [\![F, v]\!]$.

Algorithm 1 AutomataForConstraint($F \equiv C_1 \wedge C_2 \wedge \ldots \wedge C_n$)

1: **for** $v \in V_F$ **do**
2:     $\mathcal{A}(F, v) := \mathcal{A}(\Sigma^*)$;
3: **end for**
4: $i := 0$; *done* := **false**;
5: **while** $i < bound \wedge \neg done$ **do**
6:     **for each** $C \in F$ and $v \in V_C$ **do**
7:         construct A' where $\mathcal{L}(A') = \mathcal{L}(\mathcal{A}(F, v)) \cap \mathcal{L}(\mathcal{A}(C, v))$;
8:         $\mathcal{A}(F, v) := A'$;
9:     **end for**
10:     **if** none of the $\mathcal{L}(\mathcal{A}(F, v))$ changed during the current iteration of the while loop **then**
11:         *done* = **true**;
12:     **end if**
13:     $i = i + 1$;
14: **end while**

7.2.1.4   Automata Construction for Relational Constraints

For constraints that are not pseudo-relational, we extend the above algorithm to compute an over approximation of $[\![F, v]\!]$. In relational constraints, more than one variable can be involved in multi-variable constraints which creates a cycle in constraint evaluation.

Given a relational constraint in the form $F \equiv C_1 \wedge C_2 \wedge \ldots \wedge C_n$, we start with initializing each $\mathcal{A}(F, v)$ to $\mathcal{A}(\Sigma^*)$, i.e., initially variables are unconstrained. Then, we process each constraint as we described above to compute new automata for the variables in that constraint using the automata that are already available for each variable. We can stop this process at any time, and, for each variable v, we would get an over-approximation of the truth-set $\mathcal{A}(F, v) \supseteq [\![F, v]\!]$. We can state this algorithm as in Algorithm 1.

In order to improve the efficiency of the above algorithm, we first build a constraint dependency graph where, (1) a multi-variable constraint C_i depends on a single variable constraint C_j if $V_{C_j} \subseteq V_{C_i}$, and (2) a multi-variable constraint C_i depends on a multi-variable constraint C_j if $V_{C_j} \cap V_{C_i} \neq \emptyset$. We traverse the constraints based on their ordering in the dependency graph and iteratively refine the automata in case of cyclic dependencies. Note that, in the constructions we described above we only constructed automaton for the variable on the left-hand-side of a relational constraint using the automata for the variables on the right-hand-side of the constraint. In the general case we need to construct automata for variables on the right-hand-side of the relational constraints too. We do this using techniques similar to the ones we described above. Constructing automata for the right-hand-side variables is equivalent to the pre-image computations used during backward symbolic analysis as discussed in Chap. 4 [125] and we use the constructions given there. Finally, unlike pseudo-relational constraints, a relational constraint can contain negation of a basic constraint C_i where $|V_{C_i}| > 1$. In such cases, in constructing the truth set of $\neg C_i$ we can use the complement language $\Sigma^* - [\![C_i, v]\!]$ only if $[\![C_i, v]\!]$ is a singleton set. Otherwise, we construct an over approximation of the truth set of $\neg C_i$.

7.3 Relational Constraint Solving with Multi-Track DFA

In the previous section we present algorithms to construct DFA that accept satisfying values of the variables for a given formula. The algorithms we described generates a single-track DFA for each variable. Instead, we can generate a multi-track DFA that accepts tuples of variables. In this section, we discuss mapping relational string and integer constraints to multi-track DFA.

Let us extend the automata constructor function $\mathcal{A}$ such that, given a formula F and a set of variables $v_1, v_1, \ldots, v_n$, $\mathcal{A}(F, v_1, v_1, \ldots, v_n)$ generates a n-track multi-track DFA with by traversing the syntax tree of the formula (Algorithm 2). The generated multi-track DFA contains a track for each variable appearing in the formula. If a variable does not appear in the formula, the output multi-track DFA still contains a track that accepts any input for that variable, i.e., it contains an unconstrained track.

Algorithm 2 accepts input constraints of any form, e.g., DNF, CNF. Conjunctions and disjunctions are handled with $\sqcap$ and $\sqcup$ operations, respectively. Since the negation operator is not monotonic and since we sometimes over-approximate solution sets for constraints, in line 3, we convert the given formula to negation normal form (NNF) (by pushing negations to the children of boolean connectives). We provide DFA constructions for the negated constraints that over-approximate satisfying inputs to a given constraint when necessary.

Conjunction and disjunctions are handled by DFA product. The algorithm makes sure that generated DFAs have the same number of tracks by passing all variables that appear in the constraint to DFA constructor function. Next, we discuss DFA construction for string and integer constraints that do not contain boolean connectives except negation.

Algorithm 2 Multi-track DFA constructor function

1: **function** $\mathcal{A}(F, v_1, v_1, \ldots, v_n)$
2:     **if** $F \equiv \neg F$ **then**
3:         **return** $\mathcal{A}(\text{ToNegationNormalForm}(\neg F), v_1, v_1, \ldots, v_n)$
4:     **else if** $F \equiv F_1 \vee F_2$ **then**
5:         **return** $\mathcal{A}(F_1, v_1, v_1, \ldots, v_n) \sqcup \mathcal{A}(F_2, v_1, v_1, \ldots, v_n)$
6:     **else if** $F \equiv F_1 \wedge F_2$ **then**
7:         **return** $\mathcal{A}(F_1, v_1, v_1, \ldots, v_n) \sqcap \mathcal{A}(F_2, v_1, v_1, \ldots, v_n)$
8:     **else if** $F \equiv v_1 = v_2 c$ or $F \equiv v_1 \neq v_2 c$ **then**
9:         **return** $\mathcal{A}(v_1 \odot v_2 c, \Sigma^*_1, \Sigma^*_2, \ldots, \Sigma^*_n)$, where $\odot \in \{=, \neq\}$
10:     **else if** $F \equiv v_1 = cv_2$ or $F \equiv v_1 = cv_2$ **then**
11:         **return** $\mathcal{A}(v_1 = cv_2, \Sigma^*_1, \Sigma^*_2, \ldots, \Sigma^*_n)$, where $\odot \in \{=, \neq\}$
12:     **else if** $F \equiv c = v_1 v_2$ or $F \equiv c \neq v_1 v_2$ **then**
13:         **return** $\mathcal{A}(c = v_1 v_2, \Sigma^*_1, \Sigma^*_2, \ldots, \Sigma^*_n)$, where $\odot \in \{=, \neq\}$
14:     **else if** $F \equiv v_1 = v_2 v_3$ or $F \equiv v_1 \neq v_2 v_3$ **then**
15:         **return** $\mathcal{A}(v_1 = v_2 v_3, \Sigma^*_1, \Sigma^*_2, \Sigma^*_3, \ldots, \Sigma^*_n)$, where $\odot \in \{=, \neq\}$
16:     **else if** F is a linear integer arithmetic constraint **then**
17:         **return** $\mathcal{A}(F, v_1, v_1, \ldots, v_n)$
18:     **end if**
19: **end function**

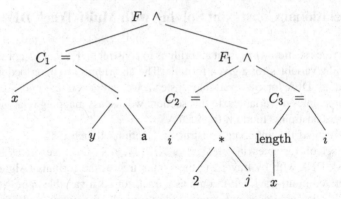

Fig. 7.5 Syntax tree of the formula $F \equiv x = ya \wedge i = 2j \wedge \text{length}(x) = i$

Fig. 7.6 The DFA
constructed for the constraint
$C_1 \equiv x = ya$

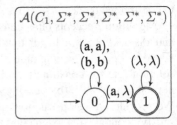

Let us consider the following example $F \equiv x = ya \wedge i = 2j \wedge \text{length}(x) = i$ where x, y are string variables, i, j are the integer variables. Constraint $C_1 \equiv x = ya$ is a relational string constraint, $C_2 \equiv i = 2j$ is a linear integer arithmetic constraint, and $C_3 \equiv \text{length}(x) = i$ is a *mixed constraint*, i.e., a constraint that contains both string and integer variables. More specifically it is an integer constraint that includes length of a string variable. To handle such constraints we introduce auxiliary variables. For the constraint C_3, we introduce an auxiliary integer variable l_v that corresponds to the length of the string variable v. In that case, we say $V_F = \{x, y, i, j, l_v\}$ is the set of variables for the constraint F. Algorithm 2 constructs multi-track DFAs by traversing the syntax tree of the constraint in Fig. 7.5.

7.3.1 Relational String Constraint Solving

DFA construction function in Algorithm 2 constructs multi-track DFA for word equations using the DFA constructions described in Chap. 5. If there are additional variables passed to the constructor function, it generates an unconstrained track for each additional variable. Consider the example constraint C_1 in Fig. 7.5 again. DFA constructor $\mathcal{A}(x = ya, \Sigma^*, \Sigma^*, \Sigma^*, \Sigma^*, \Sigma^*)$ constructs the multi-track DFA that encodes the relation between variables. Assume $\Sigma = \{a, b\}$ for string variables and $\Sigma = \{0, 1\}$ for integer variables. Figure 7.6 shows the DFA generated for the constraint C_1. For illustrative purposes, we omit to display unconstrained tracks for the multi-track DFAs. For constraint C_1 we do not show the unconstrained tracks generated for the variables i, j, and l_v.

There are also other string constraints such as begins, contains, ends and string operations such as charat, indexof, substring. The multi-track DFA construction algorithm can be expanded to handle such constraints and operations [11].

7.3.2 Relational Integer Constraint Solving

Integer automata constructor at line 17 in Algorithm 2 handles arithmetic constraints consisting of linear equalities ($=$), disequalities ($\neq$), and inequalities ($<, \leq, >, \geq$).

Given a linear integer arithmetic constraints F, function $\mathcal{A}$ first extracts the coefficients of the of the integer terms in the form $\sum_{i=1}^{n} c_i \cdot v_i + c_0 \otimes 0$ where c_i denotes integer coefficients and v_i denotes integer variables and $\otimes \in \{=, \neq, >, \geq, \leq, <\}$. Then, the automata construction techniques that rely on a binary adder state machine construction is used to construct an automaton for the arithmetic constraint [16].

Figure 7.7 shows the DFA constructed for the linear integer equality constraint C_2. Note that, we encode integer numbers as bit strings using 2's complement form.

7.3.3 Mixed String and Integer Constraint Solving

Let us consider the example mixed constraint $C_3 \equiv \text{length}(x) = i$. We rewrite the mixed constraint with auxiliary integer variable l_v as $l_v = i$ which can be solved as linear integer arithmetic constraints. The DFA constructor function for the linear integer arithmetic constraint solver converts string lengths into a binary encoded DFA and updates the equation based on the string lengths. It also generates a string DFA for the variable x where the lengths of the string variable satisfy the integer

Fig. 7.7 The DFA constructed for the constraint $C_2 \equiv i = 2j$

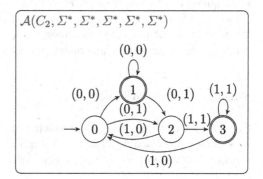

Fig. 7.8 The DFA
constructed for the constraint
$C_3 \equiv \mathrm{length}(x) = i$

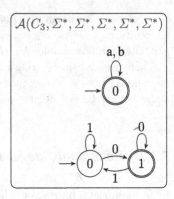

constraints [11]. As a result a multi-track DFA that contains tracks for the variables
x, i, and l_v is generated. For illustrative purposes, instead of displaying one multi-track DFA for all variables, we display a separate multi-track DFA for each variable type. Figure 7.8 shows the generated DFAs for the string and integer variables.

So far we have constructed DFAs for the leaves C_1, C_2, and C_3 of the constraint F. When handling conjunctions, we compute the values for the auxiliary variable l_v and string variable V by implementing string lengths to binary encoded integer variable conversion, and binary encoded integer variable to string variable conversion [11]. For example, the algorithm first computes a DFA for the sub constraint F_1 and computes the DFA for the formula F. When processing the conjunction for F_1, the values for the l_v and v are computed with mixed constraint handling. When processing the final conjunction for the formula F, the same procedure is applied again. Figure 7.9 shows the final DFAs constructed for the string and integer variables for the constraint F.

7.4   Model Counting

Model counting is an extension to constraint solving where instead of just answering to the question "*Does there exist a model that satisfies a given constraint?*" we try to answer the question "*How many models are there that satisfies a given constraint?*" We provide an example below to demonstrate a use case for model counting. Model counting constraint solvers are crucial tools for quantitative program analysis.

Strength of a Password Policy

Consider the example in Fig. 7.10 which is a C string manipulation example that is originally presented as a use case for model counting in [75]. On UNIX, users use the PASSWD utility to change their passwords. The example is a simplified

Fig. 7.9 The DFAs
constructed for the constraint
$F \equiv x = ya \wedge i = 2j \wedge \text{length}(x) = i$

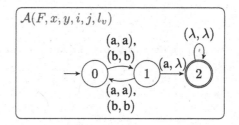

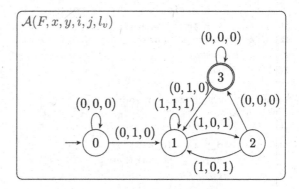

version of a C code called OBSCURE which is used by PASSWD utility to check
the password strength. At line 1, `string_checkher_helper` functions checks
if any of the parameters is a substring of the other one (`strcasestr` function
works as a case insensitive substring check). At line 8, `string_checker` function calls `string_checkher_helper` function twice; first with the original
parameters and then by reversing the input parameter p_1. If strength check fails,
`obscure_msg` function warns user for the similarity to the old password.

Suppose an attacker learns old password and the constraints imposed on new
password by OBSCURE utility. The model counting question is, how many possible
new password values are there for the attacker to try?

Let us assume old password is "abc-16" and attacker is trying to estimate the
number of all possible new passwords. The obscure function checks if one does
not contain the other or its reverse in a case insensitive manner. The example code
updates the password only if the new password is not too similar to the old one. A
symbolic execution tool can identify PCs that result in the password update, i.e., the
relation between new password and old password can be expressed in terms of a PC.
For example, the following is a path constraint that leads to the password update:

```
1  static int string_checker_helper (const char* p1, const char* p2)
        {
2    if (strcasestr(p2, p1) != NULL || strcasestr(p1, p2) {
3      return 1;
4    }
5    return 0;
6  }
7
8  static int string_checker (const char* p1, const char* p2) {
9    ...
10   int ret = string_checker_helper(p1, p2); ...
11   char* p = reverse_of(p1); ...
12   ret |= string_checker_helper(p, p2); ...
13   return ret;
14 }
15
16 static const char* obscure_msg(const char* old_p, const char*
        new_p, const struct passed* pw) {
17   ...
18   if (old_p && old_p[0] != '\0'){
19     if (string_checker(new_p, old_p)) {
20       return "similar to old password";
21     }
22   }
23   ...
24   return NULL;
25 }
```

Fig. 7.10 A string manipulation example in C language

$$strcasestr(NEW\_P, \text{"abc-16"}) = NULL \wedge$$
$$strcasestr(\text{"abc-16"}, NEW\_P) = NULL \wedge$$
$$strcasestr(NEW\_P, \text{"61-cba"}) = NULL \wedge$$
$$strcasestr(\text{"61-cba"}, NEW\_P) = NULL$$

A string model counter can count number of solutions to symbolic variable NEW_P that satisfies the given PC. This information can be used for inferring the value of the new password if an attacker knows the value of the old password. Using model counting, one can assess the likelihood of an attacker guessing the new password, hence, can evaluate the strength of the password policy.

7.4.1 Automata-Based Model Counting

In this section, we describe how to perform model counting by making use of the automata constructed by the constraint solving procedure we discussed earlier. The *model counting problem* is to determine the size of $[\![F]\!]$, which we denote $\#[\![F]\!]$. A formula can have infinitely many models. However, we can count the number of models within an infinite space of solutions restricted to a finite range for the free variables. Hence, we perform *parameterized model counting* for string constraints, in which $\#[\![F]\!](b)$ is a function over parameters b, which bounds the length of string solutions.

The constraint solving procedure described in this chapter produces a final automaton, A, for each variable in a given formula. If multi-track automaton is used, a final multi-track automaton A is constructed for a given formula. The model counting techniques we discuss here works for any DFA whether it is a single-track or multi-track. The only difference is in the interpretation of the count results; the former counts solutions to a single variable, whereas the latter counts solutions to tuples of variables. We make use of function $\#F_A(k)$ that works identical for single-track and multi-track automata.

We rely on the observation that counting the number of strings of length k in a regular language, L, is equivalent to counting the number of accepting paths of length k in the DFA that accepts L. That is, by using a DFA representation, we reduce the parameterized model counting problem to counting the number of paths of a given length in a graph. In a DFA, there is exactly one accepting path for every recognized string. Thus, there is no loss of precision due to the model counting procedure; any loss of precision comes from the over-approximations of non-regular constraints in the solving phase.

Let T be the transfer matrix of a DFA A. The matrix entry $T_{i,j}$ is the number of transitions from state s_i to state s_j. We compute $uT^k v$, where u is the row vector such that $u_i = 1$ if and only if i is the start state and 0 otherwise, and v is the column vector where $v_i = 1$ if and only if i is an accepting state and 0 otherwise. Matrix multiplication based counting method is parameterized in the following sense: after a constraint is solved, we can count the number of solutions of any desired size k by computing $uT^k v$, without re-solving the constraint.

Consider the DFA A for the constraint $F \equiv \neg\mathrm{match}(x, (01)^*)$ presented in Fig. 7.11a. Let $\mathcal{L}(A)$ ($\mathcal{L}(A) = [\![F, x]\!] = [\![F]\!]$) be the language over $\Sigma = \{0, 1\}$ that satisfies the formula F. Then $\mathcal{L}(A)$ is described by the expression $\Sigma^* - (01)^*$.

We first apply a transformation and add an extra state, s_{n+1}. The resulting automaton is a DFA A' with λ-transitions from each of the accepting states of A to s_{n+1} where λ is a new padding symbol that is not in the alphabet of A. Thus, $\mathcal{L}(A') = \mathcal{L}(A) \cdot \lambda$ and furthermore $|\mathcal{L}(A)_i| = |\mathcal{L}(A')_{i+1}|$. That is, the augmented DFA A' preserves both the language and count information of A. Recalling the final automaton from Fig. 7.11a, the corresponding augmented DFA is shown in Fig. 7.11b. (Ignore the dashed λ transition for the time being.)

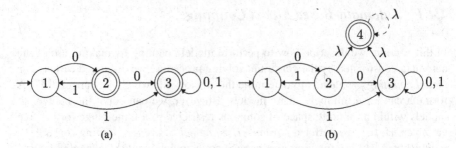

Fig. 7.11 (a) The original DFA A, and (b) the augmented DFA A' used for model counting (sink state omitted)

For our example, we show the transition matrix T and its exponentiations T^2 and T^3:

$$T = \begin{bmatrix} 0 & 1 & 0 & 0 \\ 1 & 0 & 0 & 0 \\ 1 & 1 & 2 & 0 \\ 0 & 1 & 1 & 0 \end{bmatrix}, T^2 = \begin{bmatrix} 1 & 0 & 0 & 0 \\ 0 & 1 & 0 & 0 \\ 3 & 3 & 4 & 0 \\ 2 & 1 & 2 & 0 \end{bmatrix}, T^3 = \begin{bmatrix} 0 & 1 & 0 & 0 \\ 1 & 0 & 0 & 0 \\ 7 & 7 & 8 & 0 \\ 3 & 4 & 4 & 0 \end{bmatrix}$$

Here, $T_{2,1}$ is 1 because there is a single transition to state 2 from state 1, $T_{3,3}$ is 2 because there are two transitions from state 3 to itself, $T_{4,2}$ is 1 because there is a single (λ) transition to state 4 from state 2, and so on for the remaining entries.

If we look at the entry $T_{4,1}$, we can tell the number of transitions to accepting state from start state. In other words this is equivalent to number of paths from start state to the accepting state with lengths equal to 0. The same entry on the matrix T^2 gives us the number of paths from start state to the accepting state with lengths equal to 1. We can see that $T^2_{4,1} = 2$ as there are strings "0λ" and "1λ" from the initial state to the accepting state with length 1 (omitting λ). Similarly, $T^3_{4,1} = 3$ as there strings "00λ", "10λ", and "11λ" from the initial state to the accepting state with length 2 (omitting λ).

The matrix multiplication method relies on computing $uT^k v$ and so we seek to implement an efficient method for computing this product. The time and space complexity trade-offs between various methods of computing $uT^k v$ for counting are well-studied [88, 99]. We note that one may compute T^k using matrix-matrix multiplication with successive squaring, or one may perform left-to-right vector-matrix multiplication. While successive squaring has a better worst-case time complexity bound, we found that due to typically high sparsity of DFA transfer matrices, it is both faster and less memory intensive to use repeated vector-matrix multiplication. The value of $uT^k v$ may simply be computed left to right: $uT^k v = (uT)T^{k-1}v$. This prevents us from using a divide and conquer technique, but with the benefit that at each step we are multiply a $1 \times n$ vector by a sparse $n \times n$ matrix. Hence, we need only keep track of the sparse matrix T and a single n-dimensional vector of large integers at each step. In our exploration of model counting algorithms for DFA, we have found this to be the best approach.

Another way to compute the number of paths of length k accepted by A is to employ algebraic graph theory [18] and analytic combinatorics [39] to perform model counting. A preferable solution is to derive a symbolic function that given a length bound k outputs the number of solutions within bound k. One way to achieve this is to use the *transfer matrix method* [39, 88, 99] to produce an ordinary generating function which in turn yields a linear recurrence relation that is used to count constraint solutions. Given a DFA A and length k we can compute the generating function $g_A(z)$ such that the kth Taylor series coefficient of $g_A(z)$ is equal to $|\mathcal{L}_k(A)|$ using the transfer-matrix method [39, 99].

From A' we construct the $(n+1) \times (n+1)$ transfer matrix T. A' has $n+1$ states $s_1, s_2, \ldots s_{n+1}$. Then the generating function for A is

$$g_A(z) = (-1)^n \frac{\det(I - zT : n+1, 1)}{z \det(I - zT)}, \tag{7.1}$$

where $(M : i, j)$ denotes the matrix obtained by removing the ith row and jth column from M, I is the identity matrix, $\det M$ is the matrix determinant, and n is the number of states in the original DFA A. The number of accepting paths of A with length exactly k, i.e. $|\mathcal{L}(A)_k|$, is then given by $[z^k]g_A(z)$ which can be computed through symbolic differentiation.

Applying Eq. (7.1) results in the following GF:

$$g_{A'}(z) = -\frac{\det(I - zT : n, 1)}{z \det(I - zT)} = \frac{2z - z^2}{1 - 2z - z^2 + 2z^3}. \tag{7.2}$$

7.4.2 Counting All Solutions within a Given Bound

The above described method gives a generating function that encodes each $|\mathcal{L}(A)_i|$ where i is the length bound *separately*. Instead, we seek a generating function that encodes the number of *all solutions within a bound*.

The method described above computes the number of string solutions of length *exactly* k. It is of interest to compute $\#F_A(k)$, the number of solutions *within* a given bound. This is accomplished easily by using a common "trick" that is often used to simplify graph algorithms. We add a single λ-cycle (the dashed transition in Fig. 7.11b) to the accepting state of the augmenting DFA A'. Then $\mathcal{L}(A')_{k+1} = \bigcup_{i=0}^{k} \mathcal{L}(A)_i \cdot \lambda^{k-i}$ and the accepting paths of strings in $\mathcal{L}(A')_{k+1}$ are in one-to-one correspondence with the accepting paths of strings in $\bigcup_{i=0}^{k} \mathcal{L}(A)_i$. Consequently, $|\mathcal{L}(A')_{k+1}| = \sum_{i=0}^{k} |\mathcal{L}(A)_i|$. We apply the transfer matrix method on A' to count all solutions within a given length bound.

We have shown model counting methods for counting strings of a given length. The same methods can perform model counting for any type of DFA that has different encodings, e.g., single-track string DFA, multi-track string DFA, multi-track binary encoded integer DFA [16].

7.5 Summary

In this chapter we discussed string constraint solving and model counting using automata-based techniques. We discussed automata based constraint solving techniques using single-track automata and multi-track automata. We discussed automata constructions for non-relational, relational and mixed constraints. We showed that automata-based string constraint solving leads to a model-counting string constraint solver that, given a constraint, generates (1) an automaton that accepts all solutions to the given string constraint, and (2) a model-counting function that, given a length bound, returns the number of solutions within that bound.

Chapter 8
Vulnerability Detection and Sanitization Synthesis

Web application development is error prone and results in applications that are vulnerable to attacks by malicious users. The global accessibility of Web applications makes this an extremely serious problem. According to the Open Web Application Security Project (OWASP)'s top ten list that identifies the most serious web application vulnerabilities, the top three vulnerabilities in 2007 [84] were: (1) Cross Site Scripting (XSS) and (2) Injection Flaws (such as SQL Injection). Even after it has been widely reported that web applications suffer from these vulnerabilities, XSS and SQL Injection vulnerabilities remained among the top three vulnerabilities listed in OWASP's top ten list in 2010 [85] and 2013 [86].

A *XSS vulnerability* results from the application inserting part of the user's input in the next HTML page that it renders. Once the attacker convinces a victim to click on a URL that contains malicious HTML/JavaScript code, the user's browser will then display HTML and execute JavaScript that can result in stealing of browser cookies and other sensitive data. An *SQL Injection vulnerability*, on the other hand, results from the application's use of user input in constructing database statements. The attacker can invoke the application with a malicious input that is part of an SQL command that the application executes. This permits the attacker to damage or get unauthorized access to data stored in a database.

As we stated earlier, all these vulnerabilities are caused by improper string manipulation in server-side code. Programs that propagate and use malicious user inputs with improper sanitization on the server-side are vulnerable to these well-known attacks. The attacks that exploit the vulnerabilities related to string manipulation can be characterized as *attack patterns*, i.e., regular expressions that specify potential attack strings. In this chapter we demonstrate how to use these attack patterns as security policies against which we verify and repair vulnerabilities.

© Springer International Publishing AG 2017

T. Bultan et al., *String Analysis for Software Verification and Security*,
https://doi.org/10.1007/978-3-319-68670-7_8

8.1 Vulnerability Detection and Repair

In this chapter, we present automata-based vulnerability detection and repair synthesis techniques for web applications [128]. We focus on vulnerability analysis of web applications written in PHP, however, the approach we present is applicable to all web application development languages.

The approach we present for detecting and repairing web application vulnerabilities consists of three main phases: (1) taint analysis, (2) vulnerability analysis, and (3) sanitization analysis as summarized in Fig. 8.1 [128]. Taint analysis is conducted first to find out *tainted sinks*, where input of a sensitive function depends on the value of a user input. For example, for SQL Injection vulnerabilities, a tainted sink could be a mysql_query() function whose input value depends on an external user name. As another example, for XSS vulnerabilities, a tainted sink could be an echo function whose input value depends on a user input. For each tainted sink, the taint analysis generates a dependency graph to specify how the user input flows to the sink. If there are any tainted sinks detected during taint analysis, then vulnerability analysis is conducted on the dependency graph of each tainted sink. On the other hand, if there are no tainted sinks found, the input program is considered secure.

Vulnerability analysis is carried out as a forward symbolic reachability analysis. The goal of vulnerability analysis is to determine, given an arbitrary user input, whether a string value that matches the attack pattern can reach the sink of the dependency graph. We use automata-based forward symbolic reachability analysis to compute the set of string values that can reach each node of the dependency graph. This is done by assuming that user input nodes can take any possible string value, and then propagating the post-images of string operations until a fixpoint of the automata associated with the sink has been reached. Upon termination, we take the intersection of the DFA of the sink node with the attack pattern. If the intersection is empty, we claim that the tainted sink is not vulnerable. If not, a vulnerability is reported along with a DFA that accepts the reachable attack strings (the strings that are in the intersection).

Once we have an automaton characterizing the attack strings that can reach the sink, we conduct sanitization analysis to repair the identified vulnerability. Sanitization analysis consists of vulnerability signature generation (generating a characterization of user inputs that can exploit the identified vulnerability) and patch synthesis (using the generated vulnerability signature as a filter to block or modify user inputs). A vulnerability signature is a DFA that accepts an over approximation of strings that a malicious user can provide as input that can lead to a string value at the sink that matches the attack pattern.

When a detected vulnerability depends on a single user input, we use the single-track DFA representation and backward reachability analysis to construct the vulnerability signature. When the detected vulnerability depends on multiple user inputs we use forward reachability analysis with multi-track automata to keep the relation among inputs and generate a relational vulnerability signature.

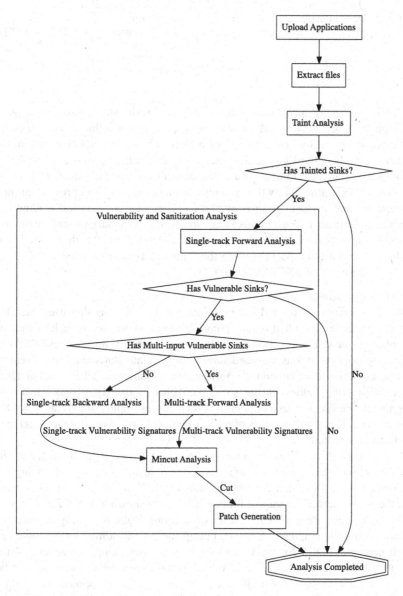

Fig. 8.1   Web application sanitization analysis process

Below we first give a simple example that has only one user input to illustrate the whole process of detecting and repairing a vulnerability with a vulnerability signature that is characterized as a single-track DFA. Then, we use another example that has two user inputs to illustrate vulnerability detection and repair with a relational signature that is characterized as a multi-track DFA.

Fig. 8.2 A simple example

```
1 <?php
2     $name = $_GET["name"];
3     $out = "NAME : " $name;
4     echo $out;
5 ?>
```

Consider the PHP script shown in Fig. 8.2. This script starts with assigning the user input provided in the _GET array to the variable name in line 2. It concatenates a constant string with variable name and assigns it to another variable out in line 3. Then it simply outputs the variable out using the echo statement in line 4.

The echo statement in line 4 is a sink statement since it can contain a Cross Site Scripting (XSS) vulnerability. For example, a malicious user can provide an input that contains the string constant <script and execute a command leading to a XSS attack. In order to prevent this vulnerability, it is necessary to *sanitize* the user inputs before using them in an echo statement. Let us assume that the attack pattern for this vulnerability is specified using the following regular expression $\Sigma^* < \Sigma^*$ (where Σ denotes any ASCII character).

Vulnerability Analysis

We first perform a forward symbolic reachability analysis that uses one DFA for each variable at each program point to represent the set of values that the variables can take. During forward analysis we iteratively update these DFAs by computing post-conditions (forward image) of program statements. For example, the post-condition computation for an assignment statement takes a set of DFAs characterizing the values of the string variables at the right-hand-side of the assignment (before the assignment is executed) as input, and returns a DFA characterizing the possible values of the left-hand-side variable after the assignment statement is executed.

During forward analysis we characterize all the user input as Σ^*, i.e., the user can provide any string as input. Any variable that is assigned an input is represented by a DFA that accepts the language Σ^* at the next program point after the assignment. For example for the small script shown in Fig. 8.2, the forward analysis will generate a DFA for the variable name at the beginning of statement 3 that accepts the language Σ^*. Computing the post-condition of the statement 3 will generate a DFA for the variable out at the beginning of statement 4 that accepts the language NAME : Σ^*. When symbolic reachability analysis reaches a fixpoint each string variable at each program point is associated with a DFA that characterizes all possible values that variable can take at that program point. This analysis is conservative in the sense that the resulting DFAs accept an over-approximation of all possible values of the variables they represent. Note that approximation is inevitable since string analysis problem is undecidable as we discussed in Chap. 2.

When the forward analysis converges, we take the intersection (using automata product) of the language of the DFA that corresponds to the string expression at the sink statement with the attack pattern. In our running example statement 3 is a sink statement, and the DFA that corresponds to the string expression at line 4 (which is simply the variable out) accepts the language NAME : Σ^*. When we take the intersection of this language with the attack pattern we obtain an automaton that accepts the language NAME : $\Sigma^* < \Sigma^*$. This automaton characterizes all possible attack strings at the sink statement. Since the language of this automaton is not empty, we know that the program is vulnerable.

Vulnerability Signature Generation

Next, we figure out which input values can create the attack strings at the sink statement. In the single-track DFA based approach, this is done with a backward symbolic reachability analysis. We start with the DFA that characterizes the attack strings (i.e, the DFA we compute at the end of the vulnerability analysis) and propagate the results backwards until we reach an input. During backward analysis we iteratively update these DFAs by computing pre-condition (backward image) of program statements. For example, the pre-condition computation for an assignment statement takes a DFA characterizing the values of the string variable at the left-hand-side of the assignment (after the assignment is executed) as input, and returns a set of DFAs characterizing the possible values of the variables that are at the right-hand-side of the assignment before the assignment statement is executed. For the example shown in Fig. 8.2, backward analysis computes the pre-condition for the assignment statement in line 3 and generates a DFA for the variable name at the end of statement 2 that accepts the language $\Sigma^* < \Sigma^*$. When we compute the pre-condition of the assignment statement in line 2 we reach an input and generate the vulnerability signature for the input _GET ["name"] as a DFA that accepts the language $\Sigma^* < \Sigma^*$.

Sanitization Generation

The last phase of the analysis generates a patch that removes the vulnerability. The vulnerability signature gives an over-approximation of all possible input values that can exploit the vulnerability. Hence, if we do not allow input values that match the vulnerability signature then we can remove the vulnerability. In the *match-and-block* strategy we generate a patch that simply checks if the input string matches the vulnerability signature. If it does, it halts the execution without executing the rest of the script. The patch generated for the small example in Fig. 8.2 based on the vulnerability signature $\Sigma^* < \Sigma^*$ and using the match-and-block strategy is shown in Fig. 8.4a. Note that the patched script will block any input string that contains the symbol $<$.

In the *match-and-sanitize* strategy, instead of blocking the execution, we modify the input in a minimal way to guarantee that the modified input cannot lead to any attack strings. We do this by analyzing the vulnerability signature DFA. Consider the DFA for the vulnerability signature $\Sigma^* < \Sigma^*$ shown in Fig. 8.3 (we use $\Sigma - \{<\}$ to indicate any symbol other than $<$). Our goal is to find a minimal set of characters, such that if we remove those characters from a given string, the resulting string will not be accepted by the DFA. As we discuss in Sect. 8.6, this corresponds to finding a cut in the graph defined by the states and the transitions of the DFA, i.e., finding a set of edges such that when we remove them, there are no paths left in the graph

Fig. 8.3 A single-track
vulnerability signature for the
example in Fig. 8.2

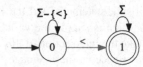

```
1       <?php
1.1       if (preg_match(
            '/([=-\xfd]|[\x00-;])*<([\x00-\xfd])*/',$_GET["name"]))
1.2       die("Invalid input");
2         $name = $_GET["name"];
3         $out = "NAME : " . $name;
4         echo $out;
5       ?>
```

(a)

```
1       <?php
1.1       if (preg_match(
            '/([=-\xfd]|[\x00-;])*<([\x00-\xfd])*/',$_GET["name"]))
1.2       $_GET["name"] =
              preg_replace('/</',"",$_GET["name"]);
2         $name = $_GET["name"];
3         $out = "NAME : " . $name;
4         echo $out;
5       ?>
```

(b)

Fig. 8.4 Patches for the example in Fig. 8.2. (**a**) Patch 1 using match-and-block strategy. (**b**) Patch 2 using match-and-sanitize strategy

from the initial state of the DFA to a final state. Note that each edge of the DFA is labeled with a symbol. After we find a cut, if we take the union of the symbols of the edges in the cut, we obtain a set of symbols such that any string accepted by the DFA must include at least one of the symbols in that set.

We use a min-cut algorithm to compute a cut that contains minimum number of edges. Then we generate a patch that deletes all the characters from the input that appear on the edges included in the cut set. For the DFA shown in Fig. 8.3, the min-cut algorithm returns the single edge labeled with the symbol < (colored in red). So we generate a patch that deletes all the < symbols from the input as shown in Fig. 8.4b. Note that, unlike the patch shown in Fig. 8.4a, the patch generated based on the match-and-sanitize strategy continues to execute the script after the sanitization.

Relational Vulnerability Signature Generation

Consider the simple script shown in Fig. 8.5. This example is similar to the one shown in Fig. 8.2 with one significant difference: there are two input variables that both contribute to the string expression used at the sink statement at line 5.

Fig. 8.5 A simple example
with two user inputs

```
1  <?php
2     $title = $_GET["title"];
3     $name = $_GET["name"];
4     $out = "NAME : " . $title . $name;
5     echo $out;
6  ?>
```

Assume that we use the single-track automata based analysis described above to analyze this script. The set of attack strings generated for the sink statement at line 5 will again be: NAME : $\Sigma^* < \Sigma^*$. However, the result of the backward analysis will be different. The crucial step is the pre-condition computation for the statement in line 4. The input to this pre-condition computation will be a DFA that accepts the attack strings characterized by the regular expression given above. The result of the pre-condition computation will generate two DFAs, one for the variable name and one for the variable title, and these DFAs will characterize all possible values these two variables can take just before the execution of statement in line 4 that can lead to generation of an attack string at the sink statement in line 5. When we do this pre-condition computation we get two DFAs that accept the same language Σ^*, i.e., any value of either variable can lead to an attack string. Although this is a sound approximation it fails to capture the information that *at least one of these variables should contain the character* $<$. Note that this condition cannot be expressed as a constraint on an individual variable, it identifies a *relation* between the two string variables.

The relational analysis we present uses a single multi-track automaton (MDFA) for each program point to capture the relationship between the input values and possible values of string expressions in the program. We use a forward analysis that operates on the dependency graph. We show the dependency graph for the example from Fig. 8.5 in Fig. 8.7. We write the string expression in the program and its corresponding string operation that corresponds to each node in the dependency graph. The analysis starts from the input nodes and traverses the dependency graph while generating one MDFA for each internal node of the dependency graph. For an input node, each MDFA has one track for each input variable and one track for the string expression that corresponds to that node, and represents the relation between them. The concatenation of two nodes results in an MDFA that has the union of input tracks of two nodes and has its string expression recorded in another track that concatenates the values of string expressions of two nodes. In Fig. 8.7 we show a string constraint on the right side of each internal node. That string constraint characterizes the set of strings accepted by the MDFA for that node. For example, for node $n3$, the string constraint is $n3 = i1.i2$ which indicates that the string expression that corresponds to node $n3$ is equal to the concatenation of input $i1$ and input $i2$.

When the analysis reaches a sink node, we intersect the track that corresponds to the string expression for the sink node (in our example this would be the track that corresponds to node $n6$) with the attack pattern DFA (by extending the attack pattern DFA to an MDFA by adding extra tracks that accept all strings). After the intersection, we project away the track for the sink node, leaving only the tracks

Fig. 8.6 A multi-track
vulnerability signature for the
example in Fig. 8.5

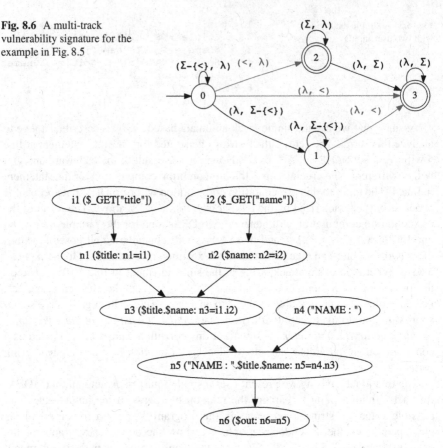

Fig. 8.7 Dependency graph

for the input nodes. The resulting MDFA represents the relational vulnerability signature. For our example, the vulnerability signature MDFA is shown in Fig. 8.6 (where each transition is marked with two symbols, one for each track, and if a track is marked with the symbol λ then that means that no symbol from that track is consumed when that transition is taken). Note that this automaton accepts tuples of strings, where either the first string in the tuple or the second string in the tuple contains at least one $<$ symbol.

The patches shown in Fig. 8.4 for the single input case are generated by converting the standard DFA representation to a regular expression and then using the PHP `preg_match` function to generate the match part of the patch. For the relational case, when we generate two regular expressions, one for each input, from the automaton shown in Fig. 8.6, we again get Σ^* for both inputs, so all inputs match. This is acceptable if we use the match-and-sanitize strategy since, although all the input strings will be considered potentially vulnerable, only a small set of symbols that relate to the vulnerability will be replaced. For example, the patch

```
1  <?php
1.1     if (preg_match('/([\x00-\xfd])*/', $_GET["title"])
        and preg_match('/([\x00-\xfd])*/', $_GET["name"])) {
1.2        $_GET["title"] =
                preg_replace('/</','',$_GET["title"]);
1.3        $_GET["name"] =
                preg_replace('/</','',$_GET["name"]); }
2       $title = $_GET["title"];
3       $name = $_GET["name"];
4       $out = "NAME : " . $title . $name;
5       echo $out;
6  ?>
```

Fig. 8.8 Patch for the example from Fig. 8.5

generated using this approach for the example in Fig. 8.5 is shown in Fig. 8.8. However, if we use the match-and-block approach using the regular expression Σ^*, we will block all the inputs which is not acceptable. As we discuss in Sect. 8.6, in such cases it is necessary to generate match statements that use automata simulation instead of automata to regular expression conversion.

In order to generate the sanitization statements from relational vulnerability signatures, we find a min-cut in the vulnerability signature MDFA as we did for the single-track case. Then, for each track, we take the union of the symbols on that track for all the edges in the min-cut. In order to sanitize the input we need to remove the symbols for each track from the input that corresponds to that track. For example, based on the min-cut shown in Fig. 8.6 (colored in red), we need to delete the symbol < both from the inputs _GET["name"] and _GET["title"]. The automatically generated replace statements for this example are shown in Fig. 8.8.

8.2 Patching Algorithm

Our patching algorithm (Algorithm 1) takes four inputs (1) the PHP web application that needs to be analyzed and fixed, (2) the attack patterns that characterize possible attacks, (3) the patching strategy to follow and (4) a specification of output functions that should be considered as sinks.

The algorithm starts (lines 1,2) by running dependency and taint analysis on the input program. Dependency analysis computes data dependencies in the application and generates dependency graphs, one graph per each specified sink. After that, taint analysis helps to identify potentially vulnerable sinks (i.e., sinks that depend on external input).

The output of these two analyses is a set of tainted dependency graphs which show how external inputs flow into potentially vulnerable sinks. Figure 8.7 shows the dependency graph for the PHP code in Fig. 8.5. Formally, a dependency graph $G = \langle N, E \rangle$ is a directed graph, where N is a finite set of nodes and $E \subseteq N \times N$

is a finite set of directed edges. An edge $(n_i, n_j) \in E$ identifies that the value of n_j depends on the value of n_i, e.g., assign the value of the variable associated with n_i to the variable associated with n_j in the program. Each node $n \in N$ can be (1) a normal node including input, constant, variable, or (2) an operation node including concat and replace.

An input node identifies the data from untrusted parties, e.g., an input from web forms. A constant node is associated with a constant value. Both nodes have no predecessors.

A concat node n has two predecessors: the prefix node $(n.p)$ and the suffix node $(n.s)$, and stores the concatenation of any value of the prefix node and any value of the suffix node in n.

A replace node n has three predecessors: the target node $(n.t)$, the match node $(n.m)$, and the replacement node $(n.r)$. For each value of $n.t$ it: (1) identifies all the matches, i.e., any value of $n.m$, that appear in $n.t$, (2) replaces all these matches in $n.t$ with any value of $n.r$, and (3) stores the result in n.

For $n \in N$, $Succ(n) = \{n' \mid (n, n') \in E\}$ is the set of successors of n. $Pred(n) = \{n' \mid (n', n) \in E\}$ is the set of predecessors of n. For a dependency graph G, we also define $Root(G) = \{n \mid Pred(n) = \emptyset\}$ and $Leaf(G) = \{n \mid Succ(n) = \emptyset\}$.

Given the extracted dependency graphs, the algorithm proceeds to the three main stages (1) vulnerability analysis, (2) signature and relational signature generation and (3) sanitization synthesis (lines 3–18). The algorithm uses two arrays $POST$ and PRE to communicate analysis results among the three stages and we will discuss these two arrays in more details later. In the following three sections, we explain the algorithms used for each of the three stages.

8.3 Vulnerability Analysis

The goal of the vulnerability analysis is to detect vulnerabilities in the input program so that they can be patched later. Given the set of tainted dependency graphs extracted from the input program, the patching algorithm (Algorithm 1 lines 5–7) runs vulnerability analysis on each of these graphs to detect if it contains a vulnerability with respect to an attack pattern.

The vulnerability analysis algorithm (Algorithm 2) takes the following inputs: a tainted dependency graph (G), an attack pattern ($attkPtrn$) specified as a regular expression and represented as a DFA, and two automata arrays $POST$ and PRE. It works by first approximating—as a regular language—the set of string values that may reach a node n in the input graph G (lines 1–5). This approximation is carried out using forward symbolic reachability analysis (Algorithm 3) which we explain later. Then, it compares the language associated with the *sink* node with the language of the *attkPtrn* (line 7–12). If the two languages intersect, this means that a vulnerability is found. In fact, the language of the intersection (i.e., language of DFA *tmp*) contains the set of reachable attack strings at the *sink* node that can be

---

Algorithm 1 PATCHER(*Prog*, *AttkPtrns*, *Strategy*, *SinkSpec*)

1: *Sinks* :=GETSINKS(*G*, *SinkSpecs*);
2: *TaintedDepGraphs* := DEPANDTAINTANALYSIS(*Prog*, *Sinks*);
3: **for** each *G* ∈ *TaintedDepGraphs* **do**
4:     array *POST*, *PRE*;
5:     **for** each *attkPtrn* ∈ *AttkPtrns* **do**
6:         *isVul* := VULANALYSIS(*G*, *attkPtrn*, *POST*, *PRE*);
7:         **if** *isVul* = true **then**
8:             *InputNodes* := GETINPUTNODES(*G*);
9:             **if** |*InputNodes*| = 1 **then**
10:                 *vulSig* := VULSIGGEN(*G*, *InputNodes*, *POST*, *PRE*);
11:             **else**
12:                 *vulSig* := RELSIGGEN(*G*, *InputNodes*, *attkPtrn*);
13:             **end if**
14:             *patch* := GENERATEPATCH(*vulSig*);
15:             report *patch*;
16:         **end if**
17:     **end for**
18: **end for**

Algorithm 2 VULANALYSIS(*G*, *attkPtrn*, *POST*, *PRE*)

1: INIT(*POST*, *PRE*);
2: **for** each *inputNode* ∈ *InputNodes*(*G*) **do**
3:     *POST*[*inputNode*] := Σ^*;
4: **end for**
5: FORWARDANALYSIS(*G*, *POST*);
6: *sink* := GETSINK(*G*);
7: *tmp*: = *POST*[*sink*] ⊓ *attkPtrn*;
8: **if** $\mathcal{L}(tmp) \neq \emptyset$ **then**
9:     *PRE*[*sink*] := *tmp*;
10:     **return** true;
11: **else**
12:     **return** false;
13: **end if**

---

used to exploit the vulnerability. This language is used later in the next phases to compute bad inputs and construct a patch.

The algorithm associates each node *n* in *G* with its automata by utilizing the two automata arrays *POST* and *PRE*. *POST*[*n*] is the DFA accepting all possible values that node *n* can take. *PRE*[*n*] is the DFA accepting all possible values that node *n* can take to exploit the vulnerability. The size of both arrays is bounded by |*N*|. Initially (line 1), all these automata accept nothing, i.e., their language is empty. Then (lines 2–4), each input node is associated with Σ^* indicating that any string value can be taken as input.

We use a forward symbolic reachability analysis (Algorithm 3) based on a standard work queue algorithm. We iteratively update the automata array *POST* until a fixpoint is reached. At line 6, $\mathcal{A}(n)$ returns a DFA that: (1) accepts arbitrary strings if *n* is an input node, (2) accepts an empty string if *n* is a variable node,

---

Algorithm 3 FORWARDANALYSIS($G, POST$)

1: queue $WQ := NULL$;
2: WQ.enqueue($Root(G)$);
3: **while** $WQ \neq NULL$ **do**
4:      $n := WQ$.dequeue();
5:      **if** $n \in Root(G)$ **then**
6:          $tmp := \mathcal{A}(n)$;
7:      **else if** n is concat **then**
8:          $tmp := $ POSTCONCAT($POST[n.p], POST[n.s]$);
9:      **else if** n is replace **then**
10:         $tmp := $ POSTREPLACE($POST[n.t], POST[n.m], POST[n.r]$);
11:     **else**
12:         $tmp := \bigsqcup_{n' \in Pred(n)} POST[n']$;
13:     **end if**
14:     $tmp := (tmp \sqcup POST[n]) \nabla POST[n]$;
15:     **if** $tmp \not\subseteq POST[n]$ **then**
16:         $POST[n] := tmp$;
17:         WQ.enqueue($Succ(n)$);
18:     **end if**
19: **end while**

---

or (3) accepts the constant value if n is a constant node. At lines 8 and 10, we incorporate two automata-based string manipulating functions that we have defined in Chap. 4:

- POSTCONCAT(DFA A_1, DFA A_2) returns a DFA A that accepts $\{w_1 w_2 \mid w_1 \in \mathcal{L}(A_1), w_2 \in \mathcal{L}(A_2)\}$.
- POSTREPLACE(DFA A_1, DFA A_2, DFA A_3) returns a DFA A that accepts $\{w_1 c_1 w_2 c_2 \ldots w_k c_k w_{k+1} \mid k > 0, w_1 x_1 w_2 x_2 \ldots w_k x_k w_{k+1} \in \mathcal{L}(A_1), \forall_i, x_i \in \mathcal{L}(A_2), w_i$ does not contain any substring accepted by $A_2, c_i \in \mathcal{L}(A_3)\}$.

At line 14, we incorporate the automata widening operator ∇ that we have defined in Chap. 6 to accelerate the fixpoint computation, which ensures termination and returns the *least* fixpoint under certain conditions. Upon termination of the while loop (lines 3 to 19) $POST[n]$ records the DFA whose language includes all possible string values that n can take.

8.4   Vulnerability Signatures

When our patching algorithm (Algorithm 1) finds a vulnerability, the next step is to compute the vulnerability signature (lines 7–13). The vulnerability signature is the set of input values that can be used to exploit a discovered vulnerability and it is represented in our analysis using a DFA. Depending on the number of inputs that flow into a vulnerable sink in a dependency graph (line 9), there are two vulnerability signature generation algorithms: (1) a non-relational vulnerability

---

Algorithm 4 VULSIGGEN($G, In, POST, PRE$)

1: $n := $ GETINPUTNODE(In);
2: $sink := $ GETSINK(G);
3: $path := $ GETPATH($G, n, sink$);
4: $PRE := $ BACKWARDANALYSIS($G, path, sink, POST, PRE$);
5: **return** $PRE[n]$;

---

signature generation algorithm (Algorithm 4) which is used when there is only a single input that can flow into the vulnerable sink, and (2) a relational vulnerability signature generation algorithm (Algorithm 6) which is used when there is more than one input that can flow into the vulnerable sink.

In this section we present our non-relational vulnerability signature generation algorithm (Algorithm 4) and leave the other one for the next section. Algorithm 4 uses a backward symbolic reachability computation based on single-track DFAs (line 4). It starts by finding a path between each pair of an input node n and a vulnerable sink s (lines 1–3). This is done to optimize the analysis since, during the backward analysis, we only need to process the nodes on this path. Then, it calls backward analysis algorithm (Algorithm 5) to compute the vulnerability signature.

Given a vulnerable sink s, backward analysis starts the computation using the input value $PRE[s]$. Recall that, during the vulnerability analysis phase, $PRE[s]$ is set to the intersection of $POST[s]$ and $attkPtrn$. Similar to the forward analysis, the computation is based on a standard work queue algorithm.

We first put the predecessors of s into the work queue (lines 2–6). We iteratively update the PRE array (by adding pre-images) until we reach a fixpoint. If the successor of n is an operation node, the pre-image (tmp) of n is computed in lines 13, 15 and 19 by calling the defined automata-based functions: PRECONCATPREFIX, PRECONCATSUFFIX, and PREREPLACE which we define in Chap. 4. Otherwise, the pre-image of n is directly derived from the successor of n (line 22). Note that $POST[n]$ records all possible values that n can take. We use this information during the pre-image computation by restricting the arguments of operations such as replace. We union the pre-images of n as tmp' at line 24.

Since we are interested only in reachable values of n, i.e., $PRE[n] \subseteq POST[n]$ by definition, we intersect tmp' with $POST[n]$ at line 26. Similar to the forward analysis, we widen the result at line 27 to accelerate the fixpoint computation. At line 28, we intersect tmp' with $POST[n]$ again to remove unreachable values (that might have been introduced due to widening) at node n. If tmp' accepts more values than $PRE[n]$, we update $PRE[n]$ at line 30 and add the predecessors of n to the working queue at line 31. Upon termination, $PRE[n]$—where n is the input node—records the DFA that accepts all possible values of n that may exploit the identified vulnerability.

Algorithm 5 BACKWARDANALYSIS(G, *path*, *sink*, *POST*, *PRE*)

1: queue $WQ = NULL$;
2: **for** each $n \in Pred(sink)$ **do**
3:      **if** $n \in path$ **then**
4:          WQ.enqueue();
5:      **end if**
6: **end for**
7: **while** $WQ \neq NULL$ **do**
8:      $n := WQ$.dequeue();
9:      $tmp' := NULL$;
10:     **for** each $n' \in Succ(n)$ **do**
11:         **if** n' is concat **then**
12:             **if** n is $n'.l$ **then**
13:                 $tmp :=$ PRECONCATPREFIX($PRE[n']$, $POST[n'.r]$);
14:             **else**
15:                 $tmp :=$ PRECONCATSUFFIX($PRE[n']$, $POST[n'.l]$);
16:             **end if**
17:         **else if** n' is replace **then**
18:             **if** n is $n'.t$ **then**
19:                 $tmp :=$ PREREPLACE($PRE[n']$, $POST[n'.m]$, $POST[n'.r]$);
20:             **end if**
21:         **else**
22:             $tmp := PRE[n']$;
23:         **end if**
24:         $tmp' := tmp' \sqcup tmp$;
25:     **end for**
26:     $tmp' := tmp' \sqcap POST[n]$;
27:     $tmp' := (tmp' \sqcup PRE[n]) \nabla PRE[n]$;
28:     $tmp' := tmp' \sqcap POST[n]$;
29:     **if** $tmp' \not\subseteq PRE[n]$ **then**
30:         $PRE[n] := tmp'$;
31:         **for** each $n' \in Pred(n)$ **do**
32:             **if** $n' \in path$ **then**
33:                 WQ.enqueue(n');
34:             **end if**
35:         **end for**
36:     **end if**
37: **end while**
38: **return** PRE;

8.5 Relational Signatures

For a vulnerable dependency graph G with multiple inputs contributing to the vulnerability, our patching algorithm (line 12) calls the relational signature generation algorithm (Algorithm 6) to generate a relational vulnerability signature. A relational vulnerability signature A of n inputs is a MDFA over the n-track alphabet Σ^n, defined as $(\Sigma \times \{\lambda\}) \times \ldots \times (\Sigma \times \{\lambda\})$ (n times), where $\lambda \notin \Sigma$ is the special symbol for padding. We further restrict A, so that all tracks are aligned and for any $w \in \mathcal{L}(A)$, $w[i] \in \lambda^* \Sigma^* \lambda^*$ ($1 \leq i \leq n$). Let $w'[i]$ denote the longest λ-free substring of $w[i]$.

Algorithm 6 RELSIGGEN(G, In, $attkPtrn$)

1: INIT($MPOST$, G, In);
2: $sink$:= GETSINK(G);
3: queue WQ := $NULL$;
4: **for** $n \in In \cup Root(G)$ **do**
5:     WQ.enqueue($Succ(n)$);
6: **end for**
7: **while** $WQ \neq NULL$ **do**
8:     n := WQ.dequeue();
9:     **if** n is concat **then**
10:         A : = CONCATSIGNATURE($MPOST[n.p]$, $MPOST[n.s]$);
11:     **else**
12:         A : = $\bigsqcup_{n' \in Pred(n)} MPOST[n']$;
13:     **end if**
14:     A := $(A \sqcup [n]) \nabla MPOST[n]$;
15:     **if** $A \nsubseteq MPOST[n]$ **then**
16:         $MPOST[n]$:= A;
17:         WQ.enqueue($Succ(n)$);
18:     **end if**
19: **end while**
20: A := $MPOST[sink] \sqcap$ EXTEND($attkPtrn$, $|In|$);
21: Project the output track away from A;
22: **return** A;

Given a dependency graph G, a set of input nodes In, a sink node $sink$, and an attack pattern $attkPtrn$, our goal is to generate a relational vulnerability signature A such that: (1) A is an $|In|$-track MDFA. Each track is associated with an input variable X_n, $n \in In$. (2) For any word w ($w[i] \in \lambda^* \Sigma^* \lambda^*$), we have $w \in \mathcal{L}(A)$ if the following condition holds: if we set $w'[i]$ as the initial value of the input node i and propagate the values of the nodes along with G accordingly, the value of the node $sink$ matches the pattern $attkPtrn$. I.e., w identifies the malicious inputs whose combination may exploit the vulnerability.

The algorithm to generate a relational vulnerability signature is shown in Algorithm 6. We perform *forward* fixpoint computation on the dependency graph where replace nodes are ignored. Our relational vulnerability signature algorithm is not capable of handling replace statements. However, since we run the vulnerability signature generation after a vulnerability is detected, we argue that it is reasonable to ignore the sanitization statements in the code (which is the typical use for the replace statements). After we generate the relational vulnerability signature, the existing sanitization statements can be commented out and replaced with the automatically generated sanitization statements.

Similar to the other analyses we presented, we use a standard work queue algorithm incorporating the automata widening operator. Each node is associated with a signature, an $i+1$-track MDFA where the first i tracks are associated with some input variables, e.g., X_n, $n \in In$, and the last track (output track) is associated with X_o used to represent the values of the current node. More specifically, i ($0 \leq i \leq |In|$) is the number of the input variables whose values have been used to construct the values of the current node. We use a MDFA vector $MPOST$ where $MPOST[n]$ is the signature associated with node n and it specifies the relations among the values of the input variables and the values of n.

Initially, for each input node $n \in In$, $MPOST[n]$ is a 2-track MDFA (associated with X_n and X_o) that accepts the identity relation on X_n and X_o, i.e., the value of the current node is equal to the value of the input variable X_n. For a node $n \in Root(G) \setminus In$, $MPOST[n]$ is a single-track DFA (associated with X_o) that either accepts Σ^* if n is a `variable` node, or accepts a constant value if n is a `constant` node. I.e., the current value of the node is an arbitrary string or a constant. In both cases, it is not related to any input variable. For the rest, i.e., $n \notin Root(G)$, $MPOST[n]$ accepts an empty set.

After we initialize $MPOST$ at line 1, we perform the fixpoint computation. Between lines 7 and 19, we iteratively update the signature at each node until the queue is empty (reaching a fixpoint). To deal with the union or widening operator on A_1 and A_2 that may be associated with the different sets of input variables, say $\mathbf{X_1}$ and $\mathbf{X_2}$, we extend both tracks to $\mathbf{X_1} \cup \mathbf{X_2}$ and X_o by padding λs in the added tracks. We then apply standard union or widening to these extended MDFAs.

Below we describe how to concatenate two signatures: CONCATSIGNATURE(A_1, A_2), where A_1 is the signature of the prefix node and A_2 is the signature of the suffix node. Let $A_1 = \langle Q_1, \Sigma_1, \delta_1, I_1, F_1 \rangle$ be a MDFA whose tracks are associated with the set of input variables $\mathbf{X_1}$ and X_o where $\Sigma_1 = (\Sigma \cup \lambda)^{|\mathbf{X_1}|} \times \Sigma$. Let A_2 be a MDFA whose tracks are associated with the set of input variables $\mathbf{X_2}$ and X_o where $\Sigma_2 = (\Sigma \cup \lambda)^{|\mathbf{X_2}|} \times \Sigma$. We first extend A_1 and A_2 to two MDFA A_1^λ and A_2^λ that are associated with $\mathbf{X_1} \cup \mathbf{X_2}$ and X_o. We extend A_1 (prefix) to A_1^λ by adding λ in the added tracks, while we extend A_2 (suffix) to A_2^λ by adding λ in both the added tracks and the common tracks that are also associated with A_1. CONCATSIGNATURE(A_1, A_2) returns the $(|\mathbf{X_1} \cup \mathbf{X_2}| + 1)$-track MDFA that accepts the concatenation of A_1^λ and A_2^λ.

As for an example, lets consider the signature A_1 that has \$title as its input variable (Fig. 8.9a). A_1 specifies that the value of the output track (out) is equal to the value of the input track. Lets consider A_2 as another signature that specifies equality but has \$name as its input variable (Fig. 8.9b). Both have two tracks, one for the input track and one for the output track. We have A_1^λ (Fig. 8.9c) and A_2^λ (Fig. 8.9d) as the extended automata of A_1 and A_2, respectively. Both have three tracks: two input tracks for \$title and \$name, and one for the output track. Note that the added track is filled with the padding symbol λ. Figure 8.9e shows the result of the concatenation of these two signatures. The result also has three tracks. For an accepted word, we have the first track value equal to the output track (circulate in state 0) and then the second track value equal to the output track (move from state 0 to 1 and then stay in state 1).

After reaching a fixpoint, at line 20, we intersect the signature of $sink$ with the attack pattern on the output track. This is done by the standard intersection of $MPOST[sink]$ and the multi-track extension of $attkPtrn$. EXTEND($attkPtrn$, $|In|$) returns an $|In|+1$-track MDFA that accepts $\{w \mid w[X_o] \in attkPtrn\}$.

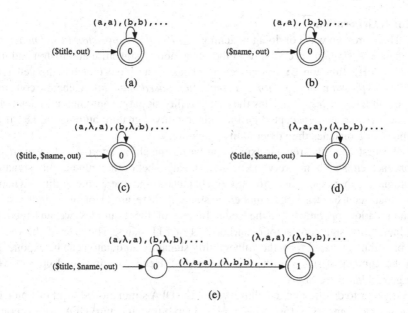

Fig. 8.9 Signature concatenation. (**a**) A_1. (**b**) A_2. (**c**) A_1^λ. (**d**) A_2^λ. (**e**) CONCATSIGNATURE(A_1, A_2)

Algorithm 7 GENERATEPATCH(*vulSig*)

1: **if** Strategy = **match-and-block then**
2:     *patch* := GENERATEBLOCKINGSIMULATOR(*vulSig*);
3: **else**
4:     Σ_{mc} := MINCUT(*vulSig*);
5:     *patch* := GENERATESANITIZER(Σ_{mc});
6: **end if**
7: **return** *patch*;

After the intersection, the output track identifies the reachable attack strings, and the input tracks identify all the malicious inputs whose combination can yield an attack string. At line 21, we project away the output track from A, and return the result at line 22 as the relational vulnerability signature of $\langle G, In, attkPtrn \rangle$.

8.6 Sanitization Generation

In this section we describe final phase of Algorithm 1 (line 14) in which we generate sanitization statements given a vulnerability signature that is characterized either as a standard single-track automaton (DFA) or a multi-track automaton (MDFA). As shown in Algorithm 7, we use two sanitization strategies: match-and-block (lines 1–3) and match-and-sanitize (lines 4–6).

In order to implement the match-and-block and match-and-sanitize strategies we need to generate code for the *match* and *replace* statements.

Match Generation

There are two ways of doing matching: (1) *Regular-expression-based matching:* Generate a regular expression from the vulnerability signature automaton and then use the PHP function `preg_match` to check if the input matches the generated regular expression, or (2) *Automata-simulation-based matching:* Generate code that, given an input string, simulates the vulnerability signature automaton to determine if the input string is accepted by the vulnerability signature automaton, i.e., if the input string matches the vulnerability signature.

We first tried the regular-expression-based matching approach. However, this approach ends up being very inefficient. The alphabet of the vulnerability signature automata consists of the 256 ASCII characters and the vulnerability signature automata can have a large number of states if there are a lot of complex string manipulation operations in the code. In one of the examples we analyzed the vulnerability signature automaton consists of 811 states. The size of the regular expression generated from the vulnerability signature automaton can be exponential in the number of states of the automaton [56]. Hence, we may end up with very large regular expressions.

In order to do efficient matching we use the DFA simulation algorithm which has linear time complexity [56]. Given the vulnerability signature DFA, we generate a function that takes a string as input, simulates the DFA, and returns true if the DFA accepts the string or false otherwise (line 2). We insert the match function instead of the `preg_match` statements shown in the patches in Figs. 8.4 and 8.8.

For the relational vulnerability signatures, we use a similar approach. Given a relational vulnerability signature characterized as an MDFA, we generate code that simulates the MDFA during the match generation. The MDFA simulation algorithm is similar to the DFA simulation algorithm, it just keeps a separate pointer for each input string to keep track of how much of each track is processed at any given time and advances the state of the MDFA based on the tuples of input symbols and the transition relation of the MDFA. The simulation time for MDFA is linear in the total length of the input strings.

Replace Generation

For the match-and-sanitize strategy, our automated sanitization generation algorithm takes the vulnerability signature automaton as input, and it generates a replace statement that modifies a given input string in such a way that the modified string is not accepted by the vulnerability signature automaton (meaning that the modified string cannot cause an attack). We modify the input strings by just deleting a set of characters using the `preg_replace` function (our approach can be extended so that escape characters can be inserted in front of a set of characters rather than deleting them). In order to prevent extensive modification to the input, the set of characters to be deleted should be as small as possible. The question is how can we identify the set of characters to be deleted?

First, we will formalize this problem in automata-theoretic terms. Let $A = \langle Q, \Sigma, \delta, q_0, F \rangle$ denote a DFA where Q is the set of states, Σ is the alphabet, $\delta \subseteq Q \times \Sigma \times Q$ is the transition relation, $q_0 \in Q$ is the initial state, and $F \subseteq Q$

is the set of accepting states. $\mathcal{L}(A)$ denotes the language accepted by A. We say $S \subseteq \Sigma$ is an *alphabet-cut* of A, if $\mathcal{L}(A) \cap \mathcal{L}_{\bar{S}} = \emptyset$, where $\mathcal{L}_{\bar{S}} = (\Sigma - S)^*$ is the set of all strings that do not contain any character in S. The *min-alphabet-cut* problem is finding the alphabet-cut S_{min}, such that for any other alphabet-cut S, $|S_{min}| \leq |S|$. For the example automaton in Fig. 8.3 the min-alphabet-cut is $\{<\}$.

The min-alphabet-cut problem can also be stated in graph-theoretic terms. Given a DFA A, an *edge-cut* of A is a set of transitions $E \subseteq \delta$ such that, if the set of transitions in E are removed from the transition relation δ, then none of the states in F are reachable from the initial state q_0. Let S_E denote the set of symbols of the transitions in E. If E is an *edge-cut* of A then S_E is an *alphabet-cut* of A. Hence, finding the min-alphabet-cut is equivalent to finding an edge-cut with minimum set of distinct symbols. For the example automaton in Fig. 8.3, the min-edge-cut is $\{(0, <, 1)\}$, which also corresponds to the min-alphabet-cut.

Note that, if the vulnerability signature DFA accepts the empty string, then there will not be any edge (or alphabet) cut since the initial state would be an accepting state. For the rest of our discussion we will assume that the DFA for the vulnerability signature does not accept the empty string (we can easily handle the cases where it accepts the empty string by first testing if the input string is empty and then inserting a single character to the input if it is).

The min-alphabet-cut problem is NP-hard [128]. This can be proven by a reduction from the vertex cover problem. A vertex cover of a graph $G = (V, E)$ is a set of vertices such that each edge of the graph is incident to at least one vertex of the set. The problem of finding a minimum vertex cover is known to be NP-complete. Vertex cover problem can be reduced to the *min-alphabet-cut* problem as follows. Given $G = (V, E)$ we build an automaton $A = \langle Q, \Sigma, \delta, q_0, F \rangle$ with the set of states $Q = E \cup \{q_0, q_F\}$, the initial state q_0, set of final states $F = \{q_F\}$, alphabet $\Sigma = V$, and the transition relation δ defined as follows: $e = (v, v') \in E \Rightarrow (q_0, v, e) \in \delta \wedge (e, v', q_F) \in \delta$. The *min-alphabet-cut* for the automaton A is the minimum vertex cover for the graph G.

Since the min-alphabet-cut problem is intractable, rather than trying to find the optimum solution, we can consider using efficient heuristics that give a reasonably small cut that is not necessarily the optimum solution. In fact, there is a very good candidate for a heuristic solution. Given a DFA A, a *min-edge-cut* of A is an edge-cut E_{min} such that for any other edge-cut E, $|E_{min}| \leq |E|$. Note that the min-edge-cut minimizes the number of edges in the edge-cut whereas the min-alphabet-cut minimizes the set of symbols on the edges in the edge-cut. Interestingly, even though the min-alphabet-cut problem is intractable, there is an efficient algorithm for computing the min-edge-cut. We use the Ford-Fulkerson's max-flow min-cut algorithm [29] to find a min-edge-cut E_{min} (line 4) where the complexity of the algorithm is $O(|\delta|^2)$. Note that $|S_{min}| \leq |E_{min}|$, i.e., the min-edge-cut provides and upper bound for the min-alphabet-cut. So if the min-edge-cut is small, then the set of distinct symbols on the edges of the min-edge-cut will give us a good approximation of the S_{min}.

Once we compute an alphabet-cut S using our heuristic, we generate a `preg_replace` statement that deletes the symbols in S from the input, making sure that the resulting string does not match the vulnerability signature (line 5).

The definition of the min-alphabet-cut problem is different for multi-track automata. Given an n-track MDFA A over $(\Sigma \cup \lambda)^n$, we say an n-tuple $S = (S_1, \ldots S_n)$, where $S_i \subseteq \Sigma$, is an alphabet-cut of A, if $\mathcal{L}(A) \cap \mathcal{L}_{\bar{S}} = \emptyset$, where $\mathcal{L}_{\bar{S}} = (((\Sigma - S_1) \cup \lambda) \times \ldots ((\Sigma - S_n) \cup \lambda)))^*$ is the set of all strings whose ith track does not contain any character in S_i. Let $|S| = |S_1| + \ldots + |S_n|$. The *min-alphabet-cut* problem for a MDFA A is finding the alphabet cut S_{min} of A, such that for any alphabet cut S of A, $|S_{min}| \le |S|$.

Since min-alphabet-cut is intractable for single-track DFA, it is also intractable for MDFA. We use min-edge-cut also as an approximation for min-alphabet-cut for MDFA. When we find a min-edge-cut, we compute the corresponding multi-track alphabet-cut by computing a set of symbols for each track by collecting the set of distinct symbols (other than λ) on each track on the edges in the min-edge-cut (line 4). The resulting alphabet cut is an n-tuple $S = (S_1, \ldots, S_n)$, where each S_i is the set of symbols for track i, i.e., input i.

Once we compute the alphabet-cuts, we generate one `preg_replace` statement for each input variable i, that deletes every symbol in S_i from the input i so that the resulting input strings do not match the vulnerability signature (line 5).

8.7 Summary

In this chapter we discussed how to detect web application vulnerabilities and how to synthesize sanitization code that repairs them via string analysis. Specifically, we discussed techniques that generate a characterization of inputs that can exploit an identified vulnerability (called vulnerability signature) and we discussed how to generate sanitization statements that eliminate the identified vulnerability. Since many critical security vulnerabilities in web applications are caused by inadequate manipulation of input strings, and given the prevalence of erroneous input validation and sanitization in web applications, it would be valuable to have an approach that automatically generates provably correct sanitizers. The technique we presented is code-sensitive, i.e., it takes into account how the application code manipulates the input value before it reaches a sink. By synthesizing customized repairs for web applications, the techniques we discuss provide a sound approach to ensuring that the detected vulnerabilities (with respect to the given attack patterns) have been eliminated.

Chapter 9
Differential String Analysis and Repair

Effectiveness of policy-based bug detection and repair that we presented in previous chapter depends on the correctness and precision of the written policies in characterizing good and bad string values. It is often possible, for instance, to encode well-known attacks into security policies (in the form of attack patterns) and write down policies for common input fields such as email address and zip code. In other cases, however, the checks to be performed on the inputs are specific to the functionality of the web application, and the input validation may be tightly coupled with and dependent on the application logic. Because they are specific to individual applications, there are no pre-specified policies that can be used to assess these types of input checks. In these cases, to make sure that the input validation is adequate, it would be necessary to specify a different policy for each different application, which is a tedious and error-prone task.

In this chapter, we present a differential analysis and repair approach [5, 7] for analyzing and repairing validation and sanitization functions in web applications. This new approach eliminates the need to write manual specifications by exploiting redundancy in input validation and sanitization code.

Web application developers often introduce redundant input validation and sanitization code in the client and server-side code of a web application. The checks done on the client-side improve the responsiveness of the application by preventing unnecessary communication with the server and reduce the server load at the same time. However, since a malicious user can by-pass the client-side checks, it is necessary to re-validate and re-sanitize at the server-side. Moreover, many applications repeat the checks for different types of fields in different parts of the application which can be exploited to obtain multiple instances of the validation and sanitization code with the same intended functionality. Finally, across different applications, one can easily find multiple instances of validation and sanitization code used to check standard formats (such as email) or to protect against same class of vulnerabilities (such as SQL injection and XSS). Using the differential analysis and repair techniques presented in this chapter, we exploit these redundancies

© Springer International Publishing AG 2017
T. Bultan et al., *String Analysis for Software Verification and Security*,
https://doi.org/10.1007/978-3-319-68670-7_9

```
1    <html>
2    ...
3    <script>
4    function validateEmail(form) {
5      var emailStr = form["email"].value;
6      if(emailStr.length == 0) {
7        return true;
8      }
9      var r1 = new RegExp("( )|(@.*@)|(@\\.)");
10     var r2 = new RegExp("^[\\w]+@([\\w]+\\.[\\w]{2,4})$");
11     if(!r1.test(emailStr) && r2.test(emailStr)) {
12       return true;
13     }
14     return false;
15   }
16   </script>
17   ...
18   <form name="subscribeForm" action="/Unsubscribe"
19       onsubmit="return validateEmail(this);">
20     Email: <input type="text" name="email" size="64" />
21     <input type="submit" value="Unsubscribe" />
22   </form>
23   ...
24   </html>
```

Fig. 9.1 JavaScript and HTML code snippets for client-side validation

within and application and across applications, to automatically detect and repair differences between input validation and sanitization functions by comparing them against each other.

A Motivating Example

Let us take a look at this example taken from a real web application called JGOSSIP (http://sourceforge.net/projects/jgossipforum/), a message board written using Java technology. Figures 9.1 and 9.2 show two snippets of client- and server-side validation code, respectively, from this application (we slightly simplified the code to make it more readable and self-contained).[1] The user fills the client-side form, shown on lines 18–22 of Fig. 9.1, by providing an email address to the HTML input element with name "email" and by clicking on the "Submit" button. When this button is clicked, the browser invokes the JavaScript function `validateEmail`, which is assigned to the `onsubmit` event of the form. This function first fetches the email address supplied by the user from the corresponding form field. It then checks if this address has zero length and, if so, accepts the empty address on line 6. Otherwise, on lines 9 and 10, the function creates two regular expressions. The

---

[1]We present the original functions rather than the IVSL extracted sanitizers to show an example of an actual difference between two validation functions written in different languages in a web application.

```
1    public class Validator {
2      public boolean validateEmail(Object bean, Field f, ..) {
3        String val = ValidatorUtils.getValueAsString(bean, f);
4        Perl5Util u = new Perl5Util();
5        if (!(val == null || val.trim().length == 0)) {
6          if ((!u.match("/( )|(@.*@)|(@\\.)/", val))
7            && u.match("/^[\\w]+@([\\w]+\\.[\\w]{2,4})$/",
8                val)) {
9            return true;
10         } else {
11           return false;
12         }
13       }
14       return true;
15     }
16   ...
17   }
```

Fig. 9.2 Java server-side validation code snippet

first one specifies three patterns that the email address should not match: a single
space character, a string with the @ symbol on both ends, and the string "@.". The
second one specifies a pattern that the email address should match: start with a set
of alphanumeric characters, followed by symbol @, further followed by another set
of alphanumeric characters, and finally terminated by a dot followed by two to four
additional alphanumeric characters. If the email address does not match the first
regular expression and matches the second one, this function returns true, indicating
acceptance of the email address (line 12), and the form data is sent to the server.
Otherwise, the function rejects the email address by returning false on line 14. This
results in an alert message to inform the user that the email provided is invalid.

When the form data is received by the server, it is first passed to the server-side
validation function. For the specific form in this example, the validation function
used is method `validateEmail` from class `Validator`, which is shown in
Fig. 9.2. This method calls a routine on line 3 to extract the value contained in the
email field from the form object (`bean`) and stores it in variable `val`. It then uses
library `Perl5Util` to perform the regular expression match operations, which
allows for using the same Perl style regular expression syntax used in the client.
First, the method checks whether the email string is `null` or has zero length after
applying the `trim` function, on line 5. If so, it accepts the string. Otherwise, it
checks the address using the same regular expressions used on the client side. As
shown on lines 6–12, the address is accepted if it satisfies these regular expression
checks, and it is used for further processing on the server side (e.g., it may be sent
as a query string to the database); otherwise, it is rejected on the server side, and the
user is taken back to the form.

As shown in this example, the regular expression checks are similar on both
ends, which emphasizes that validations on both ends should allow or reject the
same set of inputs. Otherwise, there would be mismatches that may create problems

for the application. If the server side is less strict than the client side, this would be considered a vulnerability (even when such a vulnerability is not exploitable) since it violates a common security policy that server-side checks should not be weaker than the client-side checks: a malicious user could bypass the client-side checks and submit to the server an address that does not comply with the required format, which may result in an attack. For example, an attacker could inject SQL code in the email that may result in an SQL injection attack [42]. In general, server-side checks that are less strict than the client-side checks could lead to two types of undesirable behaviors: (1) the server side allows some wrong or malicious data to enter the system, leading to failures or attacks; (2) the client side rejects legitimate values that should be accepted, resulting in the user being unable to access some of the functionality provided by the web application.

In our example, the client-side validation code shown in Fig. 9.1 rejects a sequence of one of more white space characters (e.g., " "), for which the condition on line 6 evaluates to false and the regular expression check on line 11 fails, thereby resulting in the function returning false. However, for the same input, the second condition on line 5 of the server-side validation method (Fig. 9.2) evaluates to false, due to the `trim` function call, and the string is therefore accepted by the server. This would lead to white spaces being accepted as email addresses by the server, which might in turn lead to failures (e.g., the web application might try to send an email to the user, which would fail due to an invalid email address) or attacks, such as a denial-of-service attack.

9.1   Formal Modeling of Validation and Sanitization Functions

In this section we formally specify what we mean by input *validation* and *sanitization* functions. Input validation and sanitization operations in web applications can be characterized using three types of functions: (1) *pure validator*, (2) *pure sanitizer* and (3) *validating-sanitizer* functions [5]. Each of these three types of functions can further be characterized as either a single-input or multi-input functions. We first define the single-input version of each of the three function types then generalize the definition to multi-input functions.

Pure Validators

A single-input pure validator is a total function:

$$F_v : \Sigma^* \to \{\bot, \top\}$$

that takes a string $s \in \Sigma^*$ and returns either $\top$ indicating that the string is valid and should be accepted or $\bot$ indicating the string is not valid and should be rejected.

```
1 function validateEmail(inputField, helpText){
2     if (!/.+/.test(inputField.value)) {
3          return false;
4     }
5     else {
6         if (
             !/^[a-zA-Z0-9\.-_\+]+@[a-zA-Z0-9-]+(\.[a-zA-Z0-9]
             {2,3})+$/.test(inputField.value)) {
7              return false;
8         }
9     else {
10             return true;
11         }
12    }
13 }
```

Fig. 9.3 An example of a JavaScript pure validator

A multi-input pure validator is a total function:

$$F_v : (\Sigma^*)^n \to \{\bot, \top\}$$

that takes a tuple of strings $(s_1, s_2, \ldots, s_n) \in (\Sigma^*)^n$ and returns either $\top$ indicating that all these strings are valid and should be accepted or $\bot$ indicating one of the strings s_i is not valid and hence the tuple $(s_1, s_2, \ldots, s_n)$ should be rejected.

Note that, a pure validator does not change the value of the input string, it either accepts or rejects it as it is. Figure 9.3 shows a JavaScript single-input pure validator that validates email addresses. The function makes sure that the email address is not empty (line 2) and that it matches the regular expression for valid email addresses (line 6). If these two conditions are satisfied then it accepts the input by returning true (line 10) otherwise it rejects it by returning false (lines 3,7). Notice that the email address value is not modified by the function.

Pure Sanitizers

A single-input pure sanitizer is a total function:

$$F_s : \Sigma^* \to \Sigma^*$$

that maps an input string $s \in \Sigma^*$ to an output string $s' \in \Sigma^*$.

A multi-input pure sanitizer is a total function:

$$F_s : (\Sigma^*)^n \to \Sigma^*$$

that maps an input tuple of strings $(s_1, s_2, \ldots, s_n) \in (\Sigma^*)^n$ to an output string $s' \in \Sigma^*$.

Note that, a pure sanitizer does not reject any input string, however, it may modify some of the input strings. Figure 9.4 shows a PHP single-input pure sanitizer

Fig. 9.4 An example of a
PHP pure sanitizer function

```
1  function escape($x){
2    $x = preg_replace('/"/', '\"', $x);
3    return $x;
4  }
```

Fig. 9.5 An example of a
PHP validating sanitizer
function

```
1  function reference_function($x){
2    if (strlen($x) > 4)
3      exit();
4    else {
5      $x = preg_replace('/</', '', $x);
6      if ($x == '')
7        exit();
8      else
9        return $x;
10   }
11 }
```

function. The function modifies its input by escaping each " character with a \
character (line 2) then it returns the new modified value. Notice that the function
does not reject any invalid input that contains the character ".

Validating Sanitizers

A single-input validating-sanitizer is a function:

$$F_{vs} : \Sigma^* \to \{\bot\} \cup \Sigma^*$$

that takes an input string $s \in \Sigma^*$ and either returns $\bot$ indicating that s is invalid or
maps s to output string $s' \in \Sigma^*$.

A multi-input validating-sanitizer is a function:

$$F_{vs} : (\Sigma^*)^n \to \{\bot\} \cup \Sigma^*$$

that takes a tuple of strings $(s_1, s_2, \ldots, s_n) \in (\Sigma^*)^n$ and either returns $\bot$ indicating
that one or more of the string values s_i is invalid or maps $(s_1, s_2, \ldots, s_n)$ to output
string $s' \in \Sigma^*$ by modifying and/or combining one or more of the components s_i of
the input tuple.

Note that, a validating-sanitizer may reject some inputs and modify some others.
For the rest of the dissertation we call a validating-sanitizer function a sanitizer for
short. We model all input validation and sanitization operations in web applications
as sanitizers. Note that, one can simulate a pure validator using a sanitizer: If an
input is rejected by the validator, it is rejected by the sanitizer and if it is accepted
by the validator it is returned without modification by the sanitizer. Obviously, any
pure sanitizer is also a sanitizer that never rejects an input. Hence, by just focusing
on sanitizers we are able to analyze all three types of behavior.

Figure 9.5 shows an example of a PHP single input validating-sanitizer function.
The function validates the length of the input on line 2. Then, it sanitizes the input by
deleting the character < on line 5. Finally, the function validates the result again to
make sure it is not empty on line 6. This shows how input validation and sanitization
operations are mixed.

Input Validation vs. Sanitization

Some examples of validation operations that are used in practice are PHP function `preg_match`, JavaScript function `indexof` and Java function `contains` which are utilized usually through branch conditions without modifying the values of string variables. Examples of sanitization operations are JavaScript and Java `replace` functions and PHP functions `trim`, `addslashes` and `htmlspecialchars`. These sanitization operations are typically used to update the string variables.

In web applications, there is typically a relationship between data read and write operations and the use of either input validation or input sanitization. In case of a data read operation that will not change the backend database, input sanitization can be used in order to convert potentially malicious user inputs into benign ones. This should not affect the database since these sanitized values will only be used to query the database but not to change its state.

In case of a data write operation that will change the backend database, the use of validation vs. sanitization depends on weather or not the input value is used later to query the database. If the value is going to be used later to query the database, then input validation is used to make sure that the input is in a correct format that matches the format of the data type expected by the database. For example, when signing up in a website, input fields such as *username* are usually validated only and not sanitized. The reason is that, if a sanitizer modifies a *username* value during sign up without the user's knowledge, then the user may not be able to use the original value s/he signed up with to login. Preventing attack strings that may come through these fields is done by validation operations. On the other hand, input fields for contents, such as messages in a forum, are sanitized even when they are entered into the database since they are not used to query the database later on.

Composing Sanitizer Functions

Sanitizer functions can be composed together to produce a new sanitizer function. This maybe necessary in practice if different types of attacks are expected and different sanitization functions are used to prevent them. However, as we discuss later in this chapter, composition of sanitizer functions can have subtle effects that may lead to sanitization errors.

Here we will consider the composition of single-input sanitizers only. We formally define the sanitizer composition as follows: Given two single-input sanitizer functions F_1 and F_2, their composition, $F_1 \circ F_2 : \Sigma^* \to \Sigma^* \cup \{\perp\}$, is a sanitizer function defined as:

$$F_1 \circ F_2(x) = \begin{cases} \perp & \text{if } F_2 = \perp \\ F_1(F_2(x)) & \text{if } F_2(x) \neq \perp \end{cases}$$

9.1.1 Post- and Pre-Image of a Sanitizer

In this section we discuss computing the post-image (i.e., post-condition) or pre-image (i.e., pre-condition) of a sanitizer function. In our discussion of symbolic reachability analysis in Chap. 4, we discussed computing pre and post-images of string operations and statements. In this section we generalize the pre and post-image definitions to full sanitizer functions.

Post-Image

Given an input, we call the set of strings returned by a sanitizer function its *post-image* (which is the set of strings that reach the return statement). Formally speaking, given a sanitizer F with n input variables, the set of strings returned by F when the input language for each input variable v_i is restricted to L_i where $L_i \subseteq \Sigma^*$ is defined as:

$$\text{POST}(F, (L_1, \ldots, L_n)) = \{s \mid \exists (s_1, \ldots, s_n) \in L_1 \times \cdots \times L_n : \exists s \in \Sigma^* :$$

$$F(s_1, \ldots, s_n) = s\}$$

We call this set the post-image of sanitizer F with respect to $L_1 \times \cdots \times L_n$. We can compute the post-image of a sanitizer using automata-based forward symbolic string analysis techniques discussed in Chap. 4. In general, we can not precisely compute $\text{POST}(F, (L_1, \ldots, L_n))$ due to undecidability of string analysis as we discussed in Chap. 2. Instead, we compute an over-approximation of this set, namely, $\text{POST}^+(F, (L_1, \ldots, L_n)) \supseteq \text{POST}(F, (L_1, \ldots, L_n))$. This means that, we may conclude that certain strings are accepted and returned by F when they are not. Since we are using automata-based symbolic string analysis, the result of the post-image computation is an automaton that accepts the language $\text{POST}^+(F, (L_1, \ldots, L_n))$, and we denote this automaton as $\mathcal{A}(\text{POST}^+(F, (L_1, \ldots, L_n)))$.

Figure 9.6 shows a sanitizer function F_1 along with Venn Diagrams illustrating its domain and co-domain. Function F_1 represents a single-input sanitizer function $F_1 : \Sigma^* \to \Sigma^* \cup \{\bot\}$ where $\Sigma = \{a, b\}$. Assuming Σ^* as input, the function's *post-image* $\text{POST}(F_1, \Sigma^*) = \{aa, bb, ba\}$ (notice that we always exclude $\bot$ from *post-image* as it does not represent a returned string value).

Pre-Image

Given a sanitizer function F with n number of input variables and a set of strings $L \subseteq \Sigma^*$ in the co-domain of F, we call the set of input tuples of strings that is mapped by F to L the *pre-image* of F with respect to L and we define it as:

$$\text{PRE}(F, L) = \{(s_1, \ldots, s_n) \mid \exists s \in L : F(s_1, \ldots, s_n) = s\}$$

We can use automata-based backward symbolic string analysis techniques we discussed in Chap. 4 to compute the pre-image of a sanitizer. Again, due to over-approximation, we compute the set $\text{PRE}^+(F, L) \supseteq \text{PRE}(F, L)$.

Fig. 9.6 Example of *post-image* (shaded areas) for a sanitizer function F_1 assuming input to be Σ^* where $\Sigma = \{a, b\}$

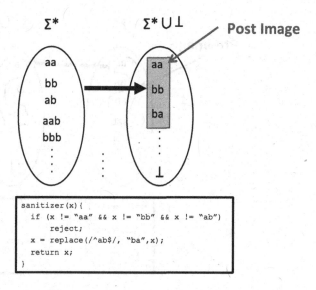

```
sanitizer(x){
    if (x != "aa" && x != "bb" && x != "ab")
        reject;
    x = replace(/^ab$/, "ba",x);
    return x;
}
```

In this chapter, we focus on non-relational analysis, i.e., we do not use relational string analysis techniques we discussed in Chap. 5. This means that for a sanitizer F with more than one input variable the pre-image computation would return the set $(\Sigma^*)^n$ which is not a useful approximation. Hence, to compute the pre-image of sanitizers with more than one input, it is necessary to use relational string analysis techniques.

Figure 9.7 shows the sanitizer function F_1 along with its *pre-image*. Given the set $\{aa, ba\}$, the *pre-image* $\mathrm{PRE}(F_1, \{aa, ba\}) = \{aa, ab\}$.

Negative Pre-Image

We call the set of strings that are rejected by a sanitizer F the *negative pre-image* of F. For a given sanitizer function F, this set is defined as:

$$\mathrm{PRE}_\perp(F) = \{(s_1, \ldots, s_n) \mid F(s_1, \ldots, s_n) = \perp\}$$

Again, we use automata-based backward symbolic string analysis techniques we discussed in Chap. 4 to compute an over approximation of the negative pre-image, $\mathrm{PRE}_\perp^+(F)$, where $\mathrm{PRE}_\perp^+(F) \supseteq \mathrm{PRE}_\perp(F)$. This means that, we may conclude that certain strings are rejected by F when they are not. On the other hand, since we are computing an over-approximation, any string that is rejected by F is guaranteed to be in $\mathrm{PRE}_\perp^+(F)$. Since we are using automata-based symbolic string analysis, the result of the negative pre-image computation is an automaton that accepts the language $\mathrm{PRE}_\perp^+(F)$, and we denote this automaton as $\mathcal{A}(\mathrm{PRE}_\perp^+(F))$.

Figure 9.8 shows the *negative pre-image* of sanitizer F_1. $\mathrm{PRE}_\perp(F_1) = \Sigma^* - \{aa, bb, ab\}$.

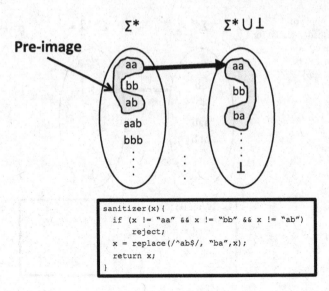

Fig. 9.7 Example of *pre-image* (the shaded area on the left) of sanitizer function F_1 given a subset of the co-domain of F_1 (shaded area on the right)

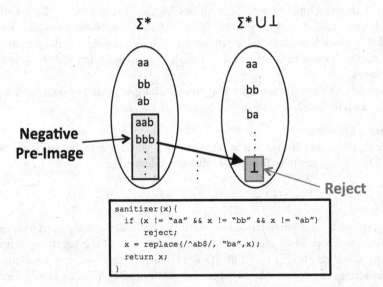

Fig. 9.8 Example of a *negative pre-image* (the shaded area on the left) of sanitizer function F_1 which is mapped by F_1 to $\perp$ (i.e., rejected)

Negative Post-Image

Negative post-image does not characterize a subset of the input or the output of a sanitizer. Given a sanitizer F and a set of possible input values L, the *negative post-image* of F with respect to L, $\text{POST}_{\perp}(F, L)$, is the set of strings that reach the negative sinks (i.e., reach the `reject` statements) in F.

As with previous image computations, in general, we can not precisely compute $\text{POST}_\perp(F, L)$ due to undecidability of string analysis. Instead we compute an over-approximation of this set, namely, $\text{POST}_\perp^+(F, L) \supseteq \text{POST}_\perp(F, L)$. This means that, we may conclude that a string can reach a negative sink when it does not. Since we are using automata-based symbolic string analysis, the result of the negative post-image computation is an automaton that accepts the language $\text{POST}_\perp^+(F, L)$, and we denote this automaton as $\mathcal{A}(\text{POST}_\perp^+(F, L))$.

9.2   Discovering Client- and Server-Side Input Validation and Sanitization Inconsistencies

In this section, we present a differential string analysis technique to discover inconsistencies between the client- and the server-side code in web applications [7]. The presented approach (1) extracts and maps input validation functions at the client and server sides, (2) models input validation functions as deterministic finite automata (DFAs) using string analysis techniques from Chap. 4, and (3) identifies and reports inconsistencies in corresponding input validation functions.

9.2.1   Extracting and Mapping Input Validation and Sanitization Functions at the Client- and the Server-Side

In order to extract validation and sanitization functions, we first need to identify entry points of the web application, that is, points where user input is read. At the client side, such points correspond to input fields in web forms. Web application frameworks typically let developers specify the input fields of a web application, and the JavaScript validation functions to be applied to each field in a configuration file. By leveraging this information, one can identify (1) the set of validated and sanitized inputs on the client side, and (2) the corresponding set of JavaScript functions that are used for validating and sanitizing such inputs.

The identification of the input validation and sanitization code on the server side is similar to that of the client side, with the difference that validation and sanitization is typically performed using a different language (for example, Java instead of JavaScript) and that parameters are read through calls to input functions. Similar to the client side, web application frameworks also allow developers to specify server side inputs and corresponding validation and sanitization functions. Therefore, an analysis of the web application's configuration file can also provide us with (1) the set of validated and sanitized inputs for the server side and (2) the set of Java functions that are used for validating and sanitizing each of those inputs.

It is worth noting that web applications could also perform input validation checks directly in the code, without explicitly specifying inputs and corresponding validating functions in a configuration file. In such cases, static and dynamic program analysis techniques can be used to extract input validation and sanitization functions [4]. One approach is to use crawling to find input fields and corresponding sinks in a web application. When the crawler hits a web page with an HTML form, it can fill it out automatically using a set of pre specified values and submit it. Then, for each HTML input field, client-side validation and sanitization code (in JavaScript) can be dynamically extracted, resulting with one sanitizer function per each input field. On the server-side, one can also collect the execution traces to figure out the inputs and the sinks (where the inputs flow into) and the code that is executed during crawling. This information can later on be used to extract the server-side sanitizer functions and to map server-side sanitizer functions to client-side sanitizer functions [4].

9.2.2   *Inconsistency Identification*

Given a client-side validator/sanitizer F_c and a server-side validator/sanitizer F_s for an HTML input field i, the *inconsistency identification* is the problem of finding if F_c and F_s return different output values for the same set of inputs. Two functions F_c and F_s are inconsistent if $\text{POST}(F_s, \Sigma^*) \neq \text{POST}(F_c, \Sigma^*)$.

Algorithm 1 identifies inconsistencies between client- and server-side validator/sanitizer functions. The algorithm takes as its input a client-side single-input validator/sanitizer F_c and a server-side single-input validator/sanitizer F_s both working on the same input field i and the type of the functions (identifying if they are validators of sanitizers). In the algorithm, each variable that has a name starting with A represents a DFA, each variable with a name starting with F represents a validator or sanitizer. The algorithm uses the DFA operations $\sqcap, \sqcup, -, \neg$ which correspond to the automata union, intersection, difference, and complement operations; $\mathcal{A}$ is the automata constructor function, and we use $\mathcal{L}$ to denote the language accepted by an automaton. The algorithm starts by computing two DFAs: $A_c(i)$ (client side) and $A_s(i)$ (server side), where $A_c(i) = \mathcal{A}(\text{POST}^+(F_c, \Sigma^*))$ and $A_s(i) = \mathcal{A}(\text{POST}^+(F_s, \Sigma^*))$ (lines 1,2).

Using $A_c(i)$ and $A_s(i)$, we construct two new automata:

- $A_{s-c}(i)$ where $A_{s-c}(i) = A_s(i) - A_c(i)$ (line 3), and
- $A_{c-s}(i)$ where $A_{c-s}(i) = A_c(i) - A_s(i)$ (line 18).

We call $A_{s-c}(i)$ and $A_{c-s}(i)$ *difference signatures,* where:

- $\mathcal{L}(A_{s-c}(i))$ contains strings that are accepted and returned by the server side but rejected by the client side, and
- $\mathcal{L}(A_{c-s}(i))$ contains strings that are accepted and returned by the client side but rejected by the server side.

Let us now consider various scenarios for the difference signatures. If $\mathcal{L}(A_{s-c}(i)) = \mathcal{L}(A_{c-s}(i)) = \emptyset$, this means that our analysis could not identify

---

Algorithm 1 INCONSISTENCYIDENTIFICATION($F_c, F_s, type$)

1: $A_c := \mathcal{A}(\text{POST}^+(F_c, \Sigma^*));$
2: $A_s := \mathcal{A}(\text{POST}^+(F_s, \Sigma^*));$
3: $A_{s-c} := A_s - A_c;$
4: **if** $\mathcal{L}(A_{s-c}) \neq \emptyset$ **then**
5:     **if** $type =$ validators **then**
6:         Find $w \in \mathcal{L}(A_{s-c});$
7:         **if** $(F_c(w) = \bot) \wedge (F_s(w) \neq \bot)$ **then**
8:             Report Bug in Server-Side Validator F_s; **return** w;   //counter example
9:         **end if**
10:    **else**
11:        $A_{s_i} := \mathcal{A}(\text{PRE}^+(F_s, A_{s-c}));$
12:        Find $w \in \mathcal{L}(A_{s_i});$
13:        **if** $((F_c(w) = \bot) \wedge (F_s(w) \neq \bot)$ **then**
14:            Report Bug in Server-Side Sanitizer F_s; **return** w;   //counter example
15:        **end if**
16:    **end if**
17: **end if**
18: $A_{c-s} := A_c - A_s;$
19: **if** $\mathcal{L}(A_{c-s}) \neq \emptyset$ **then**
20:    **if** $type =$ validators **then**
21:        Find $w \in \mathcal{L}(A_{c-s});$
22:        **if** $(F_s(w) = \bot) \wedge (F_c(w) \neq \bot)$ **then**
23:            Report Bug in Client-Side Validator F_c; **return** w;   //counter example
24:        **end if**
25:    **else**
26:        $A_{c_i} := \mathcal{A}(\text{PRE}^+(F_c, A_{c-s}));$
27:        Find $w \in \mathcal{L}(A_{c_i});$
28:        **if** $(F_s(w) = \bot) \wedge (F_c(w) \neq \bot)$ **then**
29:            Report Bug in Client-Side Sanitizer F_c; **return** w;   //counter example
30:        **end if**
31:    **end if**
32: **end if**

---

any difference between the client- and server-side validation functions, so we have no errors to report. Note that, due to over-approximation in our analysis, this does not mean that the client and server-side validation functions are proved to be equivalent. It just means that our analysis could not identify an error.

If $\mathcal{L}(A_{s-c}(i)) \neq \emptyset$, there might be an error in the server-side validation function (line 4). A server-side input validation function should not accept a string value that is rejected by the client-side input validation function—as we discussed earlier, this would be a security vulnerability that should be reported to the developer. Due to over-approximation in our analysis, however, our result could be a false positive. To prevent generating false alarms, we validate the error as follows.

We try to find an example input string that would result in the difference between the client and server-side. Since a validator does not modify its input, then we do not need to compute the preimage of the difference. Instead, we generate a string $s \in \mathcal{L}(A_{s-c}(i))$ and execute both the client and server-side input validation functions by providing s as the input value for the input field i. If client-side function rejects the string, and server-side function accepts it, then we are guaranteed that there is a problem with the application and report the string s as a counter-example to the developer. If we cannot find such a string s, then we do not report an error (lines 5–9).

If we have sanitizers then we need to do pre-image computation to get the set of input values that resulted in the difference. Then we generate the example from this set, i.e., generate $s \in \text{PRE}^+(F_s, A_{s-c})$ (lines 10–16).

We also check if $\mathcal{L}(A_{c-s}(i)) \neq \emptyset$ and, if so, we again generate a string to demonstrate the inconsistency between the client and server-side validation functions. Note that client-side validation functions accepting a value that the server rejects may not be as severe a problem as their counterpart (lines 19–32). It is nevertheless valuable to report this kind of inconsistencies because fixing them can improve the performance and response time of the web application and prevent client-side vulnerabilities [94].

9.3   Semantic Differential Repair for Input Validation and Sanitization

After finding the differences between two input validation and sanitization functions, the next step is to repair the functions against each other to remove the difference. In this section we present a semantic differential repair algorithm [5] that exploits redundancies in input validation and sanitization, within an application and across applications, to automatically repair input validation and sanitization functions by comparing them against each other.

An Example

Let us give an overview of the automated differential repair technique that strengthens the validation and sanitization functionality of a given *target* function based on a given *reference* function. Consider the example functions shown in Fig. 9.9. We are showing the original functions written in PHP language to help the reader when comparing the original functions with the generated patches.

The reference function starts with a validation check that blocks any string that is longer than 4 characters. This is followed by a sanitization operation which replaces the character < with ϵ (i.e., deletes <). Finally, the result of the sanitization operation goes through another validation check that blocks the empty string. The target function in Fig. 9.9 does not do any validation. It only sanitizes the input string by replacing the character " with the string "\ "" (i.e., it escapes the double quote characters).

The goal of the differential repair technique is to strengthen the validation and sanitization operations in the target function as much as the reference function. More precisely, the goal is to make sure that the repaired target function does not return a string that is not returned by the reference function or the original target function. Before explaining how the differential repair technique works on these

Fig. 9.9 A small, but
illustrative example, showing
a target function to be
repaired based on a reference
function

```
function reference_function($x){
    if (strlen($x) > 4)
        exit();
    else {
        $x = preg_replace('/</', '', $x);
        if ($x == '')
            exit();
        else
            return $x;
    }
}



function target_function($y){
    $y = preg_replace('/"/', '\"', $y);
    return $y;
}
```

two functions, we would like to discuss two potential repair techniques that may
seem to be the natural choice in this case and explain why they do not work. This
would help the reader to understand the motivation behind the choices that were
made in development of the differential repair algorithm.

Repair by Composition

One question we may ask is: why not simply run two sanitizers one after
another? Due to lack of idempotency in some string sanitization operations, one
can not blindly compose two given sanitizers to get a stronger one without first
computing the difference between them. For example, composing a reference and
a target sanitizers that have some differences but at the same time share the
following sanitization operation `preg_replace('/"/', '\"', $x)` —which
escapes the " with a \—is problematic. Since this operation is not idempotent,
the composition would result in double escaping i.e., "ab"c" would become
"ab\\"c" instead of "ab\"c". Furthermore, we repair sanitizers that are extracted
from different programming languages (and different applications in some cases).
The original two pieces of code where we extracted the two sanitizers from are
written in different languages with different semantics and contain other logic
related to the functionality of the application where they were extracted from.
This shows the importance of the (1) extraction phase in removing code unrelated
to validation and sanitization and (2) using a language agnostic string analysis
framework in which the semantic differences are handled by reducing them to
differences between regular languages.

The goal of repair is to make sure that the post-image of the repaired function
does not contain any string that is not in the post-image of the reference function
and the original target function.

The post-image for the reference function in Fig. 9.9 is the language of all strings
that are shorter than 5 characters and not empty and do not contain the character
<, while the post-image for the target function is the language of all strings that do

not contain the character " unless it is preceded by the character \. For example, the string "foo" is an element in the reference function's post-image while the string "foo<" is not since it contains the < character. Also, the strings "foo" and "foo\"bar" are elements in the target function's post-image while the string "foo"bar" is not since it contains the character " without being preceded by the character \.

The differential repair algorithm we discuss below works in three phases, where each phase generates a patch-function with a specific purpose: (1) a *validation patch*, (2) a *length patch*, and (3) a *sanitization patch*. The final repair is obtained by applying the composition of all three patch-functions together as we explain below.

Validation patch

The purpose of this phase is to generate a patch that makes sure that the repaired function rejects all the inputs that are rejected by the reference function. Figure 9.10 shows the validation-patch produced in this phase of the repair algorithm for our running example. The validation patch blocks all input strings that are either empty,

```
function validation_patch($x){
   if (preg_match('/<*|[^<].{4,}|<[^<].{3,}|<<[^<].{2,}|
                   <<<[^<].+/',$x))
     exit();
   else
     return $x;
}

function length_patch($x){
  if (preg_match(
    '/"".{1,2}|".{1,2}"|.{1,2}""|"[^"]{3,3}|[^"]{3,3}"/',$x))
    exit();
  else
    return $x;
}

function sanitization_patch($x){
  $x = preg_replace('/</', "", $x);
  return $x;
}

function repaired_function($x){
  return target_function(
           sanitization_patch(
             length_patch(
               validation_patch($x)
             )
           )
         );
}
```

Fig. 9.10 The repaired function that is generated by our differential repair algorithm for the target function shown in Fig. 9.9

consist of one or more < characters or longer than 4 characters. For example, the strings "", "<", "<<<" and "<html>" will be blocked by the validation patch. On the other hand, the strings "fo" and "<a>" will not be blocked.

The validation patch blocks the inputs that generate a string that is in the post-image of the target function but not in the post-image of the reference function. Note that our algorithm is able detect that some input strings are blocked by the reference function only after being sanitized such as the string "<<<" (which is first converted to empty string by deletion of < and then blocked by the reference function). So, for this case, to make sure that the string "<<<" is not in the post-image of the repaired function, the validation patch blocks it.

Length patch

The purpose of this phase is to make sure that (1) the maximum length of the strings that are in the post-image of the repaired function is not bigger than the maximum length of the strings that are in the post-image of the reference function and (2) the minimum length of the strings that are in the post-image of the repaired function is not smaller than the minimum length of the strings that are in the post-image of the reference function.

For the reference function in our example, the minimum length is 1, since it blocks the empty string, and the maximum length is 4. On the other hand, for the target function, after the validation patch is applied, the minimum length is 1 since it also blocks the empty string, but the maximum length is not 4 but 8. The reason is that the sanitization in the target function escapes the " character so that an input string of length 4 like """"""" (which passes the validation patch) is escaped to produce the string "\"\"\"\"" at the sink, which is of length 8.

This example shows that due to the sanitization operation in the target function, we get a length difference in the post-image languages even though the validation patch has already blocked all strings longer than 4. To address this issue we generate a length patch that blocks any input string that results in a string longer than 4 characters at the target sink even if the input string itself is shorter than 4 characters. For example, the length patch blocks the string ""a"" although it has 3 characters only since it will result in the string "\"a\"" of length 5 at the sink which is longer than 4 characters. On the other hand, the string "foo" will not be blocked by the length patch since it will reach the sink as it is, 3 characters long.

Figure 9.10 shows the length patch-function for our example. Note that the function assumes that the validation patch function is applied before it so it only blocks things not blocked by the validation patch function. In Sect. 9.3.2.2 we explain how to automatically generate the length patch-function.

Sanitization patch

The purpose of this final phase is to take care of the differences that are due to sanitization operations. Our goal is to make sure that the post-image of the repaired function is a subset of the post-image of the reference function.

In our example, there is one sanitization operation in the reference function in which the character < is deleted. Even after application of the validation and length patches, this behavior would not be fully replicated at the repaired target function. Although the validation patch will prevent some strings such as "<<<" from reaching the sink at the repaired function, there are still other strings, such as "a<b" for example, that will still be in the post-image of the repaired function but not in the post-image of the reference function, since the character < gets deleted. The goal of the sanitization patch is to remedy such situations, and make sure that the sanitization operations in the target function are as strong as the sanitization operations in the reference function.

Unlike the previous two phases, the sanitization patch does not block the input strings that are found in the difference between the post-images of the target and reference functions. Instead we use the *min-cut* algorithm we discussed in Chap. 8 to generate sanitization code that will delete (or escape) certain characters in the input strings such that the difference between the two post-images is removed. Using this min-cut algorithm, our differential repair algorithm will generate the sanitization patch-function shown in Fig. 9.10. This function does not block input strings that contain the character <, but rather, deletes this character from these input strings and returns the corresponding string without that character. This repair simulates the same sanitization behavior of the reference function in the new repaired function. In Chap. 8 we explained the min-cut algorithm and how to automatically generate the sanitization patch.

Given the final sanitization phase, one might think that the first two phases are redundant. However, without the first two phases, the repair generated by our approach can become too conservative by rejecting all input strings or by deleting all characters from the input string. Dividing the repair generation to three separate phases enables us to generate a combined repair that is not overly conservative.

The final result of the differential repair algorithm for our running example is shown in Fig. 9.10. The *repaired function*, is obtained by composing the three patch-functions, in the order in which they were introduced here, with the original target function.

Using the extraction techniques we discussed earlier in this chapter, we extract one sanitizer function per input field which characterizes all the validation and sanitization operations that are used for that particular field. Validation and sanitization operations involve use of regular expressions and validation operations such as string `match`, `substring`, and sanitization operations such as string `replace`, `trim`, `addslashes`, `htmlspecialchars`, etc.

9.3.1   Differential Repair Problem

Given a target sanitizer function F_T and a reference sanitizer function F_R, the goal of differential repair is to generate a new sanitizer function F^P, called a patch, such that when F_T is patched by composing it with F^P, the resulting repaired function returns a string only if F_R and F_T can both return that string. I.e., we want to make sure that a string is not in the post-image of the repaired function if it is not in the post-image of F_T or F_R.

In order to formalize this, let us define the difference between the post-images of two sanitizer functions F_1 and F_2 as follows:

$$\mathrm{DIFF}(F_1, F_2) = \{x \mid \exists y \in \Sigma^* : F_1(y) = x \wedge (\forall z \in \Sigma^* : F_2(z) \neq x)\}$$

which is the set of strings that are in the post-image of F_1 but not in the post-image of F_2. Given this definition (along with the definition of sanitizer's composition from Sect. 9.1), the differential repair problem is to automatically construct a patch F^P such that $\mathrm{DIFF}(F_T \circ F^P, F_R) = \emptyset$, which means when we compose F_T with F^P we want to make sure that the result, $F_T \circ F^P$, is at least as strict as F_R, i.e., its post-image is a subset of the post-image of F_R. We call this new composed function the *differential repair* F_{DR}, where $F_{DR} = F_T \circ F^P$.

Note that, due to the way we are constructing the differential repair, by composing the target function F_T with the automatically generated patch F^P, we guarantee that the repaired function F_{DR} is at least as strict as F_T, i.e., its post-image is also a subset of the post-image of F_T.

9.3.2 Differential Repair Algorithm

Given a target sanitizer F_T and a reference sanitizer F_R, our differential repair algorithm consists of three phases that produce three patches: (1) The *validation patch generation* phase produces F^V, (2) the *length patch generation* phase produces F^L, and (3) the *sanitization patch generation* phase produces F^S. The result of our differential repair algorithm is a patch that is the composition of these three individual patches: $F^S \circ F^L \circ F^V$ and the repair we generate is the composition of this patch with the target function, i.e., $F_{DR} = F_T \circ F^S \circ F^L \circ F^V$.

Differential repair algorithm is shown in Algorithm 2. The algorithm takes a target sanitizer F_T and a reference sanitizer F_R as input and generates sanitizer F_{DR} as output which corresponds to differential repair of F_T with respect to F_R. The differential repair algorithm computes post or pre-images of given sanitizers as DFA using the automata based string analysis techniques we discussed in previous chapters. As before, the DFA operations $\sqcap, \sqcup, -, \neg$ correspond to the automata union, intersection, difference, and complement operations; $\mathcal{A}$ is the automata constructor function, and we use $\mathcal{L}$ to denote the language accepted by an automaton. In the remaining part of this section we discuss the three phases of the Algorithm 2.

9.3.2.1 Phase I: Validation Patch Generation

Our goal is to generate a validation patch F^V such that:

$$\forall x \in \Sigma^* : F_R(x) = \bot \Rightarrow F_T \circ F^V(x) = \bot,$$

Algorithm 2 DIFFERENTIALREPAIR(F_T, F_R)

1: $A_1 := \mathcal{A}(\text{PRE}_\perp^+(F_R))$;
2: $A_2 := \mathcal{A}(\text{PRE}_\perp^+(F_T))$;
3: **if** $(\mathcal{L}(A_1 - A_2) \neq \emptyset)$ **then**
4: 　　$A^V := A_1 - A_2$;
5: 　　$F^V := \text{GENERATEBLOCKINGSIMULATOR}(A^V)$;
6: **else**
7: 　　$F^V := \text{IDENTITYFUNCTION}; A^V := \mathcal{A}(\emptyset)$;
8: **end if**
9: $A_1 := \mathcal{A}(\text{POST}^+(F_R, \Sigma^*))$;
10: $A_2 := \mathcal{A}(\text{POST}^+(F_T, \mathcal{L}(\neg A^V)))$;
11: $A_d = A_2 - A_1$;
12: **if** $(\mathcal{L}(A_d) \neq \emptyset)$ **then**
13: 　　**if** $(len_{min}(A_2) < len_{min}(A_1) \vee len_{max}(A_2) > len_{max}(A_1))$ **then**
14: 　　　　$A_3 := \text{RESTRICTLENGTH}(A_2, A_1)$;
15: 　　　　$A^L := \mathcal{A}(\text{PRE}^+(F_T, \mathcal{L}(A_2 - A_3)))$;
16: 　　　　$F^L := \text{GENERATEBLOCKINGSIMULATOR}(A^L)$;
17: 　　　　$A_2 := A_3$;
18: 　　**else**
19: 　　　　$F^L := \text{IDENTITYFUNCTION}$;
20: 　　**end if**
21: 　　$A_d := A_2 - A_1$;
22: 　　**if** $(\mathcal{L}(A_d) \neq \emptyset)$ **then**
23: 　　　　$A_{mc} := \mathcal{A}(\text{PRE}^+(F_T, \mathcal{L}(A_d)))$;
24: 　　　　$\Sigma_{mc} := \text{MINCUT}(A_{mc})$;
25: 　　　　$F^S := \text{GENERATESANITIZER}(\Sigma_{mc}, A_1)$;
26: 　　**else**
27: 　　　　$F^S := \text{IDENTITYFUNCTION}$;
28: 　　**end if**
29: **else**
30: 　　$F^S := F^L := \text{IDENTITYFUNCTION}$;
31: **end if**
32: $F_{DR} := F_T \circ F^S \circ F^L \circ F^V$; **return** F_{DR};

---

i.e., the validation patch F^V guarantees that $F_T \circ F^V$ does not accept inputs that F_R rejects. In order to compute the validation patch, we first need to identify the set of strings that are rejected by F_T and F_R i.e., their negative pre-images.

As we said before, it is not possible to compute the pre or post-image of a sanitizer precisely due to undecidability of string analysis problem. We use automata-based backward symbolic string analysis techniques discussed in Chap. 4 to compute an over approximation of the negative pre-image, $\text{PRE}_\perp^+(F)$, where $\text{PRE}_\perp^+(F) \supseteq \text{PRE}_\perp(F)$. This means that, we may conclude that certain strings are rejected by F when they are not. On the other hand, since we are computing an over-approximation, any string that is rejected by F is guaranteed to be in $\text{PRE}_\perp^+(F)$. Since we are using automata-based symbolic string analysis, the result of the negative pre-image computation is an automaton that accepts the language $\text{PRE}_\perp^+(F)$, and we denote this automaton as $\mathcal{A}(\text{PRE}_\perp^+(F))$.

In lines 1 and 2 of Algorithm 2 we construct two automata A_1 and A_2, that accept an over-approximation of the negative pre-images of F_T and F_R, respectively, where $\mathcal{L}(A_1) = \text{PRE}^+_\perp(F_R)$ and $\mathcal{L}(A_2) = \text{PRE}^+_\perp(F_T)$. The next step (line 3) checks if the reference function F_R rejects more input values than the target function F_T by computing the difference between negative pre-images of A_1 and A_2. If the difference is empty then F^V is assigned the identity function (line 7) which is a sanitizer function that returns the input as it is without blocking any value (i.e., it is a no-op). If the difference is not empty, the target function must be patched to reject the values rejected by the reference function. To achieve this we automatically generate a patch that rejects only the strings that are rejected by F_R but not F_T.

Note that the validation patch we generate is not sound due to over-approximation of the negative pre-image of the target function F_T. The set of strings that are in $\text{PRE}^+_\perp(F_R) \sqcap (\text{PRE}^+_\perp(F_T) - \text{PRE}_\perp(F_T))$ will not be blocked by the patch we generate, whereas they should be blocked in order to reach our precise goal. We can make the validation patch sound by blocking all the strings in $\text{PRE}^+_\perp(F_R)$ without computing the set difference with $\text{PRE}^+_\perp(F_T)$, but, that would result in generation of a validation patch in many cases even when it is not necessary. Our experiments indicate that the imprecision in our pre-image computation is not a problem in practice since for all the examples we manually checked we observe that $\text{PRE}^+_\perp(F_R) \sqcap (\text{PRE}^+_\perp(F_T) - \text{PRE}_\perp(F_T)) = \emptyset$.

Figure 9.11 shows the validation patch automaton A^V that is automatically generated for the example shown in Fig. 9.9 where Σ represents the ASCII characters. To save space we collapsed all transitions between any two states s_i and s_j into one transition t_{ij}. We annotate this transition with a set of characters $\Sigma_C \subseteq \Sigma$ such that if a character c is in Σ_C then there is a transition on c between s_i and s_j. The sink state along with transitions into and out of it are omitted.

Since our analysis represents the set of strings at each program point using DFA, we generate the patch repair function F^V based on the DFA that is computed by our analysis. The validation patch code that is generated with GENERATEBLOCK-INGSIMULATOR filters the inputs by simulating the resulting automaton A^V in Fig. 9.11 to determine if the input string is accepted by A^V. If the input string is accepted by the automaton A^V, then F^V will return $\perp$ to block the input, otherwise it will return the input string without modification.

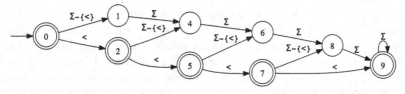

Fig. 9.11   The validation patch automaton A^V for the example in Fig. 9.9. The validation patch F^V blocks the strings accepted by this automaton

9.3.2.2 Phase II: Length Patch Generation

The goal of length patch generation is to generate a patch F^L such that:

$$\forall x \in \Sigma^* :$$
$$((\exists y, z \in \Sigma^* : |F_R(y)| \leq |F_T \circ F^V(x)| \leq |F_R(z)|) \Rightarrow F^L cal(x) = x) \land$$
$$(\neg(\exists y, z \in \Sigma^* : |F_R(y)| \leq |F_T \circ F^V(x)| \leq |F_R(z)|) \Rightarrow F^L cal(x) = \bot)$$

i.e., given the target function F_T composed with the validation patch F^V, F^L rejects any input string that will cause the output of $F_T \circ F^V$ to contain a string of length longer or shorter than all the strings in the output of the reference function F_R.

The validation patch makes sure that any input string rejected by the reference sanitizer is also rejected by the repaired target sanitizer. However, this does not mean that the set of strings that are returned by the repaired target sanitizer and the reference sanitizer are the same after the validation patch since they may be using different sanitization operations. The length patch is the first step in establishing that the repaired target sanitizer does not return any string that is not returned by the reference sanitizer. The length patch makes sure that the length of any string returned by the repaired target function is not larger or smaller than all the strings returned by the reference sanitizer.

The lines 9–20 in Algorithm 2 construct the length patch. The lines 9 and 10 compute an over-approximation of the post-images i.e., the automata that accept the set of strings that are returned by the reference sanitizer and the target sanitizer that is composed with the validation patch. The lines 11 and 12 in Algorithm 2 check if there are any strings that are returned by the target sanitizer composed with the validation patch that are not returned by the reference sanitizer by checking if $\text{POST}^+(F_T, \mathcal{L}(\neg A^V)) - \text{POST}^+(F_R, \Sigma^*) = \emptyset$. If the difference is empty, then we consider $F_T \circ F^V$ to be as strict as F_R and the analysis concludes by assigning IDENTITYFUNCTION (i.e., no-op) to length and sanitization patches F^L and F^S (line 30).

Note that, due to over-approximation in our analysis, it is not guaranteed that $F_T \circ F^V$ is as strict as F_R even if the difference is empty. However, again manual inspection of our experiments indicate that our approximate analysis always finds the differences if they exist since the precision of our post-image computation is quite good in practice.

If a difference is found, then we check if the difference corresponds to a length difference in line 13. Let us first define len_{max} and len_{min} for an automaton. Given an automaton A, $len_{max}(A) = \infty$ if A accepts an infinite set, and $len_{max}(A)$ is the length of the longest string accepted by A otherwise. We can check if $len_{max}(A) = \infty$ by checking if there are cycles in A on any path from the starting state to an accepting state. If there is at least one cycle, then $len_{max}(A) = \infty$. If there are no cycles, then $len_{max}(A)$ is finite, and we use a depth first search to compute the length of the longest string accepted by A. On the other hand, given an automaton A, $len_{min}(A)$ is

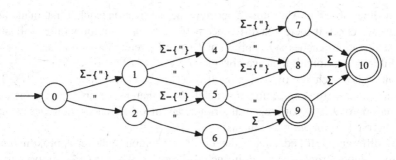

Fig. 9.12 The length patch automaton A^L for the example in Fig. 9.9. The length patch F^L rejects the strings accepted by this automaton

the length of the shortest string accepted by A. If the start state is an accepting state then $len_{min}(A) = 0$. Otherwise, $len_{min}(A)$ is computed by finding the length of the shortest path from the start state to an accepting state.

If a length difference is found, then we restrict the length of the set of strings accepted by F_T to remove the length difference using the following operation in line 14:

$$\text{RESTRICTLEN}(A_2, A_1) \equiv A_2 \sqcap \bigsqcup_{i=len_{min}(A_1)}^{len_{max}(A_1)} \Sigma^i$$

After the length restriction, in line 15, we use backward symbolic reachability analysis (that we discussed in Chap. 4) to compute an over-approximation of the set of input strings that cause the length discrepancy i.e., $\text{PRE}^+(F, L) \supseteq \text{PRE}(F, L)$. Note that, this over-approximation may result in blocking input strings that do not contribute to the length discrepancy.

In line 16 we generate the length patch F^L that blocks the strings that are accepted by the automaton A^L in Fig. 9.12 and returns the strings rejected by A^L without any change. Figure 9.12 shows the length patch automaton A^L that our algorithm computes for the example shown in Fig. 9.9.

9.3.2.3 Phase III: Sanitization Patch Generation

The third and final phase in the repair algorithm is the sanitization patch generation, which results in a patch function F^S such that:

$$\forall x \in \Sigma^* : (\forall y \in \Sigma^* : F_R(y) \neq x) \Rightarrow$$
$$(\forall z \in \Sigma^* : F_T \circ F^S \circ F^L \circ F^V(z) \neq x)$$

which means that, after adding the sanitization patch F^S to the previously generated patches, we want the differential repair $F_{DR} = F_T \circ F^S \circ F^L \circ F^V$ to be as strict as F_R in terms of the set of strings it returns.

The lines 21–28 in Algorithm 2 generate the sanitization patch. First, in the lines 21, 22, we check if there is a difference between what F_T returns (after validation and sanitization patches are applied) and what F_R returns assuming any input. If no difference is found, then we consider $F_T \circ F^L \circ F^V$ to be as strict as F_R and the analysis concludes by assigning IDENTITYFUNCTION to F^S (line 27). This indicates that there is no sanitization patch. Note that, as we discussed before, due to over-approximation our repair algorithm can miss differences, however we have not observed this in our experiments.

If a difference is found, then, in the line 23, we compute an over-approximation of the set of input strings that result in such a difference. The set $\mathcal{L}(A_{mc})$ represents an over-approximation of the set of all input strings that are the cause of the difference between the set of strings returned by F_R and $F_T \circ F^L \circ F^V$. We call A_{mc} the *mincut* automaton and in the line 24 we use this mincut automaton to generate a mincut alphabet using the approach discussed in Chap. 8, such that if the symbols in the mincut alphabet are removed from the input strings, then the difference between the post-images of F_R and $F_T \circ F^L \circ F^V$ disappear. Figure 9.13 shows the mincut automaton A_{mc} for our running example in Fig. 9.9 along with the mincut edges which correspond to the mincut alphabet $\Sigma_{mc} = \{<\}$.

Then, in the line 25, we generate the sanitization patch F^S to either delete or escape the set of symbols in the mincut alphabet. Finally, in the lines 32 and 33, we construct and return the differential repair function F_{DR} as the composition of the target function F_T with the three patch functions generated during the three phases of the repair algorithm.

Code Generation Heuristics

Once we compute an alphabet-cut Σ_{mc}, we generate the sanitization patch F^S with a `replace` statement that deletes the symbols in Σ_{mc} from the input, making

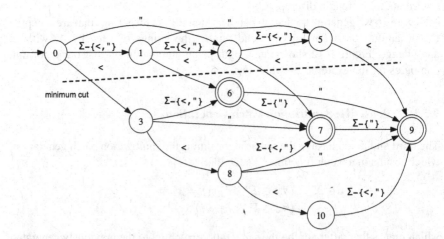

Fig. 9.13 The mincut automaton A_{mc} for the example in Fig. 9.9. The dotted line shows the mincut edges with the corresponding mincut alphabet $\{<\}$

sure that the resulting string does not match A_{mc}. Although the function F^S is a sound repair that will guarantee that $\text{POST}^+(F_T \circ F^S \circ F^L \circ F^V, \Sigma^*) \subseteq \text{POST}^+(F_R, \Sigma^*)$, we apply two heuristics to generate more accurate repair functions.

The first heuristic is the *escape* heuristic. We look at the automaton A_1 generated in line 9 of the Algorithm 2 (which represents all the string values returned by F_R), and check if all the characters in the mincut alphabet Σ_{mc} are always preceded by the same single character e. If that is the case, we call the character e the escape character. Formally speaking, given DFA $A_1 = \langle Q_1, q_0, \Sigma, \delta_1, F_R \rangle$, we check that $\forall q \in Q_1, \forall c \in \Sigma_{mc} : \delta_1(q, c) \neq sink \Rightarrow (\forall q' \in Q_1 : \delta_1(q', c') = q \Rightarrow c' = e)$. If this is the case, then the sanitization patch F^S we generate uses the `replace` operation to escape all the characters in the mincut alphabet Σ_{mc} (instead of deleting them) by prepending them with the escape character e.

The second heuristic we use is the *trim* heuristic. Here, if we get a mincut Σ_{mc} which contains space characters, we first check if A_1 accepts any string that contains a space character (which can be determined by checking if transitions on space characters always go to the sink). If not, then we generate a patch that deletes the space characters as in our basic mincut based patch generation algorithm. If A_1 does accept a string that contains a space character, then we check if it is the case that the strings accepted by A_1 do not start or end with space characters. Formally speaking, given DFA $A_1 = \langle Q_1, q_0, \Sigma, \delta_1, F_R \rangle$, we check that for all space character s $\delta_1(q_0, s) = sink$ and $\forall q \in Q_1 : \delta_1(q, c) \in F_R \Rightarrow c \neq s)$. If so, then we generate patch F^S which uses the `trim` function to delete the space characters from the beginning and end of each input string.

9.4 Summary

In this chapter we argued that the redundancy in client and server-side validation and sanitization code in web applications can be exploited to identify potential inconsistencies in validation and sanitization policies. Differential string analysis techniques can be used to identify such inconsistencies automatically. We presented a formal model for input validators and sanitizers as a basis for differential analysis. Then, we demonstrated that, automata-based symbolic string analysis techniques can be used to identify inconsistencies between input validators or sanitizers. Finally, we presented a differential repair approach that automatically strengthens an existing sanitizer to make it at least as strong as a given reference sanitizer.

Chapter 10
Tools

In this chapter we discuss several tools that implement the techniques we described in the earlier chapters. The tools we discuss are: An automata based string analysis library called LIBSTRANGER [127], a vulnerability analysis tool for PHP programs built on LIBSTRANGER called STRANGER [127], an automated repair tool for string manipulating code called SEMREP [5], and an automata-based constraint solver for string constraints called ABC [12].

10.1 LIBSTRANGER

LIBSTRANGER [127] is a string manipulation library that handles all core string and automata operations described in Chap. 4 such as general language replacement, concatenation, intersection, union, widen and special replace operations. During the string forward and backward analysis carried out by STRANGER and SEMREP, all string and automata manipulation operations that are needed to compute the post and pre-images of string operations are available in LIBSTRANGER. LIB-STRANGER uses the MONA library [24] for symbolic representation of automata using MTBDDs.

The core of LIBSTRANGER is implemented in C language as a shared library libstranger.so to get faster computation time and to have a tight control on memory. A C++/Java class called *StrangerAutomaton* is available to encapsulate the low level algorithms and data structures and provide a higher level interface to the library. We used JNA (Java Native Access) to bridge the C language and Java code. LIBSTRANGER along with its source code and manual is available at: https://github.com/vlab-cs-ucsb/LibStranger.

© Springer International Publishing AG 2017
T. Bultan et al., *String Analysis for Software Verification and Security*,
https://doi.org/10.1007/978-3-319-68670-7_10

10.2   STRANGER

STRANGER [127] (STRing AutomatoN GEneratoR) implements the techniques discussed in Chap. 8 to check the correctness of string validation and sanitization functions in Web applications with respect to known attacks.

STRANGER is implemented in Java and uses PIXY [60] as a front end and the string manipulation library LIBSTRANGER (see 10.1) along with MONA [24] for automata manipulation. STRANGER takes a PHP program and a set of attack patterns as input and automatically analyzes the given program and reports the possible vulnerabilities (such as XSS or SQL Injection characterized as attack patterns) in the program. For each input that leads to a vulnerability, it also outputs the vulnerability signature, i.e., an automaton (in a dot format) that characterizes all possible string values for this input which may exploit the vulnerability, along with the patch generated using the mincut algorithm. STRANGER and several benchmarks are available at http://www.cs.ucsb.edu/~vlab/stranger.

The architecture of STRANGER is shown in Fig. 10.1. The tool consists of mainly two modules: (1) the vulnerability analysis module that uses PIXY to parse PHP code and perform the taint analysis and then performs the vulnerability analysis and repair and (2) the string analysis module that implements the post- and pre-image computation and uses LIBSTRANGER and MONA for automata manipulation operations.

The first step in the vulnerability analysis is done by PIXY [60], a taint analysis tool for detecting web application vulnerabilities. PIXY parses the PHP program and constructs the control flow graph (CFG). PHP programs do not have a single entry point as in some other languages such as C and Java, so we process each script by itself along with all files included by that script. The CFG is passed to the taint analyzer in which alias and dependency analyses are performed to generate dependency graphs. If no tainted data flow to the sink, taint analysis reports the dependency graph to be secure; otherwise, the dependency graph is considered to be tainted and passed to the vulnerability analyzer for more inspection.

The vulnerability analyzer implements our repair algorithm (see Chap. 8).

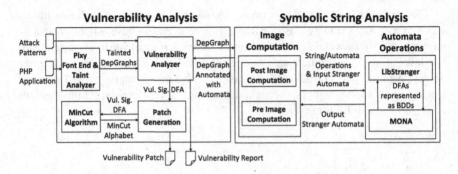

Fig. 10.1 The architecture of STRANGER

The vulnerability analyzer uses symbolic string analysis similar to post- and pre-image computation from Chap. 4 (that is modified slightly to work with dependency graphs).

The dependency graphs are pre-processed to optimize the image computation. First, a new acyclic dependency graph is built where all the nodes in a cycle (identifying cyclic dependency relations) are replaced by a single strongly connected component (SCC) node. The vulnerability analysis is conducted on the acyclic graph so that the nodes that are not in a cycle are processed only once.

In the forward analysis, we propagate the post images to nodes in topological order, initializing input nodes to DFAs accepting arbitrary strings. Upon termination, we intersect the language of the DFA of the sink node with the attack pattern. If the intersection is empty, we conclude that the sink is not vulnerable with respect to the attack pattern. Otherwise, we perform the backward analysis and propagate the pre images to nodes in the reverse topological order, initializing the sink node to a DFA that accepts the intersection of the result of the forward analysis and the attack pattern. Upon termination, the vulnerability signatures are the results of the backward analysis for each input node. For both analyses, when we hit an SCC node, we switch to a work queue fixpoint computation on nodes that are part of the SCC represented by the SCC node.

During the fixpoint computation we apply automata widening on reachable states to accelerate the convergence of the fixpoint computation. We added the ability to choose when to apply the widening operator. This option enables computation of the precise fixpoint in cases where the fixpoint computations converges after a certain number of iterations without widening. We also incorporate a coarse widening operator that guarantees the convergence to avoid potential infinite iterations of the fixpoint computation.

10.3 SEMREP

SEMREP is a tool for analysis and repair of validation and sanitization functions in web applications. SEMREP implements a language-agnostic automated semantic differential repair algorithm from Chap. 9 to analyze and repair validation and sanitization functions in web applications. Most of SEMREP is implemented in C++. MinCut algorithm and patch generator are implemented in Java. SEMREP uses the LIBSTRANGER library along with MONA library for automata manipulation operations. Source code for latest version along with the manual are available online from https://github.com/vlab-cs-ucsb/SemRep.

SemRep consists of two modules: (1) the differential repair module which implements the differential repair algorithm in Chap. 9 and (2) the symbolic string analysis module which computes the pre and post-images of a sanitizer. Figure 10.2 shows the architecture of the tool.

The tool takes as input the dependency graphs (see Chap. 8) of two sanitizer functions. After parsing the dependency graphs, difference computation component

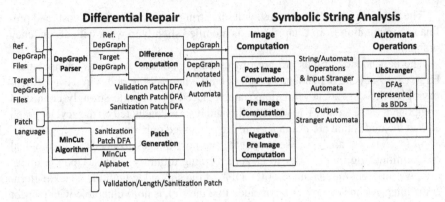

Fig. 10.2 The architecture of SEMREP

will send each graph to negative pre-image computation component. In general, image computation components use forward and backward string analyses to compute post and pre-images of the function represented by the dependency graph similar to the way done in STRANGER. Each node in the dependency graph will be annotated with a DFA (stored in an object of type `StrangerAutomaton`) that accepts all possible string values that may reach this node going forward or backward. (Negative) Pre-image component annotates the `input` node with the (negative) pre-image DFA while post-image component annotates the `return` node with the post-image DFA.

After negative pre-image computation component finishes, the difference computation component uses the two DFAs associated with the `input` nodes to calculate the validation patch. Next, it will annotate the `input` node in the target dependency graph with the validation patch DFA (if a validation patch is needed) and send the two dependency graphs to the post-image computation component. Then, it checks to find out if there is a length difference between the validation-patched target and the reference by checking the length difference between post-image DFAs that are associated with the `return` nodes. If a difference is found, it will (1) restrict the length of the DFA at the target `return` node by the length of the DFA at the reference `return` node and (2) annotate the `return` node at the target with the new length-restricted DFA indicating that the language of this DFA represents the preferred output. It then sends the target dependency graph to pre-image computation component to compute the pre-image for this preferred output which represents the length patch DFA.

Then, if after the length restriction there is still a difference between the DFAs at the sinks (`return` nodes), the difference computation component will annotate the `return` node of the target dependency graph with the DFA that represents this difference (which we call sanitization difference DFA), indicating a non-preferred output, and sends that dependency graph to the pre-image computation component. The pre-image computation component annotates the `input` node with the sanitization patch DFA. Then, the difference computation component sends the sanitization patch DFA along with the validation and length patch DFAs to the patch generation component.

The patch generation component will do the following: (1) Generate the code for the validation and length patches in the preferred programming language provided in the patch language input. These patches are functions that simulate the validation and length-patch DFAs. (2) Send the sanitization patch DFA to the mincut algorithm and uses the returned mincut alphabet to generate the code for the sanitization patch.

10.4 ABC

In this section we discuss the tool ABC (Automata-Based model Counter) that implements the automata construction and model counting techniques for string constraints. ABC is implemented based on the algorithms described in Chap. 7. ABC source code is available online at: https://github.com/vlab-cs-ucsb/SemRep.

Figure 10.3 represents the high-level architecture of ABC. We can divide ABC into two main components: (1) A compilation module which performs syntactic operations, (2) automata constructor module for constraint solving and model counting.

ABC aims to supports SMTLIB language specification as an input language in order to support different types of symbolic execution tools. However, there is no standard language syntax for string theory in SMTLIB specifications. Hence, ABC

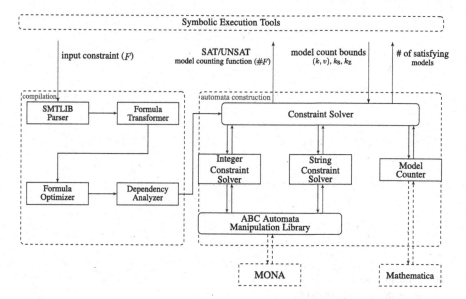

Fig. 10.3 ABC architecture

follows the syntax that CVC4 SMT solver uses for the string theory [73]. Once an input constraint is given to ABC, it first parses it and *Formula Transformer* pushes negations down into the boolean connectives. It also checks for syntactic level optimizations that can be done such as constant folding and constant propagation. Next, *Formula Optimizer* optimizes the input formula based on equivalence relations. It also generates implied numeric constraints for the string constraints that have non-regular truth sets. Next, *Dependency Analyzer* checks for the dependencies between variables and divides the input constraint into groups that does not share any common variable. At the end of compilation phase an Abstract Syntax Tree (AST) of the input constraint along with the additional information on variables are passed to automata-construction module. *Constraint Solver* is responsible for managing automata construction for different theories. *Integer Constraint Solver* and *String Constraint Solver* modules implements the algorithms described in Chap. 7. ABC also provides an automata manipulation library that models string operations from different programming languages. ABC automata manipulation library is extended from LIBSTRANGER[1] library and depends on MONA[2] automata manipulation library. *Model Counter* implements the automata-based model counting algorithms described in Chap. 7. ABC provides support for matrix exponentiation based model counting using big number libraries available to C++ and symbolic model counting using MATHEMATICA.[3]

ABC is implemented using the C++ programming language. ABC is implemented as an *autotools* project to support portability among different systems. It can be installed as an executable or as a C++ shared library. ABC also provides a JNI interface which makes it easily available for JAVA applications.

---

[1] https://github.com/vlab-cs-ucsb/LibStranger.

[2] https://github.com/cs-au-dk/MONA.

[3] https://www.wolfram.com/mathematica/.

Chapter 11
A Brief Survey of Related Work

In this chapter we provide a brief survey of related research work. We first give an overview of alternative approaches to string analysis, followed by a discussion on recent work on string constraint solvers. We discuss application of string analysis and string constraint solving techniques to bug and vulnerability detection in web applications. We conclude the section with a discussion in differential analysis and program repair techniques.

11.1 String Analysis

As we stated before, string analysis refers to static and dynamic techniques for reasoning about string expressions in a program. The goal of string analysis is to compute possible values that string expressions can take during the execution of the program. String analysis is necessary for verification of programs that use extensive string manipulation. It is especially crucial for vulnerability analysis in web applications. Below we discuss different technical approaches for string analysis.

11.1.1 Grammar Based String Analysis

JSA [26] is a static string analyzer for Java programs that has been very influential. Given a Java program, JSA first constructs a directed flow graph for every specified hotspot that captures the flow of strings and string operations while abstracting away the rest of the program. Nodes of the flow graph represent string constants, variables, expressions, and operations while edges represent possible data flows. Hotspots represent the program points where the user of the tool is interested in finding out

© Springer International Publishing AG 2017

T. Bultan et al., *String Analysis for Software Verification and Security*,
https://doi.org/10.1007/978-3-319-68670-7_11

the string values that can reach them. This flow graph is then transformed into a context free grammar (CFG) such that each nonterminal represents a node in the graph and each terminal represents a constant string. The grammar has the property that for a node n in the flow graph, the associated nonterminal A_n has $\mathcal{L}(A_n)$ (the set of strings that can be derived from nonterminal A_n) that contains all possible string values the string variable or expression represented by that node may take during program execution. This grammar is then over-approximated (using Mohri-Nederhof algorithm) using a finite state automaton A such that $\mathcal{L}(A)$ (language accepted by the automaton A) represents an over approximation of string values that may reach the hotspot.

JSA has been applied to statically analyzing the XML transformations in Java programs [66] by using DTD schemas as types and modeling the effects of XML transformation operations. Grammar-based string analysis technique has also been used to check for errors in dynamically generated SQL query strings in Java-based web applications [118]. Grammar-based string analysis technique has also been applied to analyzing executable programs for the x86 architecture [27].

Grammar-based string analysis approach has been extended by providing support for string-based replacement operations in PHP applications [78]. In this extension the whole HTML output of a PHP program is characterized all at once instead of one hotspot at a time. Instead of converting the resulting CFG into an automaton and then giving the results represented as an automaton, this technique stops after the CFG creation phase and directly uses the resulting CFG in two ways. First, it checks XSS attacks by intersecting the resulting CFG language with a regular expression language that represents the dangerous output that contains an XSS attack. Second, it checks for HTML output well-formedness by checking if the CFG language is a subset of HTML CFG language. Since this problem is undecidable, it bounds the depth of tag-nesting in HTML CFG language which converts it into a regular language. If the subset check succeeds then program's output is valid otherwise the bound is increased and a test is done again. To model PHP string operations such as *str_replace*, finite state transducers are used where a transducer transforms a CFG language based on a string operation by computing its post image under that operation.

The string analyzer presented in [78] was later used to check for SQL injection vulnerabilities in PHP applications [119]. First, a CFG language that approximates strings that may reach an SQL hotspot is computed. Nonterminals in this CFG are annotated with taint values from taint analysis. Then for each tainted nonterminal, two checks are applied. First check determines if it is in a literal string syntactic position in an SQL query. If so, then the second check determines if it has a non-empty intersection with a regular language that represents strings with odd number of un-escaped quotes. If so then it is considered to be vulnerable. A similar approach to check for XSS vulnerabilities is used in [120].

11.1.2 Automata-Based String Analysis

A symbolic automata representation is proposed in [52] and finite state transducers based on this representation are used to analyze behaviors of sanitization operations in [109, 113]. The resulting tool, called Bek, is able to identify whether a target string is a valid output of a sanitization routine. Later on, this line of research was extended to string encoding and decoding operations [32, 33, 38, 108]. These tools limit their analysis to single-input sanitizers.

The automata-based symbolic reachability analysis techniques we discuss in Chap. 4 also use a symbolic automata representation but involve computation of fixpoints to determine reachable string values rather than modeling the sanitizers directly using transducers [129, 130].

Rex [110, 111] combines SMT solving (using Z3 [76]) with symbolic automata and is effective in encoding and manipulating strings having large alphabets such as Unicode. This is a related approach to the MTBDD based symbolic automata encoding we discuss in Chap. 4 [129, 130].

As we discussed in Chap. 5, automata representation can be generalized to multi-track automata, and multi-track automata can be used to model relations between string variables [133, 134]. Rather than using multi-track automata to encode the relationship between input and output of a sanitizer, in this approach, multi-track automata is used to keep track of the relationships among the variables in a given program state.

Language-based replacement has been discussed in computational linguistics [45, 61, 80, 107]. These algorithms are based on the composition of finite state transducers. By composing specific transducers, constraints like longest match and first match can be precisely modeled. The transducer-based replacement function [80] has been implemented in Finite State Automata utilities (FSA) [106], where automata are stored and manipulated using an explicit representation.

A widening method to analyze strings has been investigated in [25]. The widening operator is defined on strings and the widening of a set of strings is achieved by applying the widening operator pairwise to each string pair. The widening operator we discuss in Chap. 6 is defined on automata, and was originally proposed in the context of automata representation of arithmetic constraints [17].

A path- and index- sensitive string analysis based on Monadic Second-Order Logic (M2L) [51] has been proposed in [103]. This technique statically encodes string operations that are used in java sanitization code into M2L and then checks if a string generated by the sanitization code satisfies a pre-specified constraint using an M2L constraint solver such as MONA [24].

11.1.3   Hybrid String Analysis

Static analysis techniques typically suffer from lack of scalability or loss of precision, or both. One approach is to use hybrid techniques that combine static analysis techniques with dynamic techniques in order to improve both precision and scalability.

In AMNESIA [50], SQL Injection attacks are fought by first applying static string analysis to approximate the syntactic structure of an SQL query at a hotspot in a program. Then dynamic monitoring is used to enforce this structure when executing the program. The key insights behind this approach is that information needed to predict the possible structure of SQL queries in a web application is contained within the application's code. So an SQLI attack, by injecting additional SQL statements into a query, would violate that structure. AMNESIA uses the static string analysis tool JSA [26] to analyze the application code and automatically build a model for the legitimate queries. This analysis is applied per each hotspot in the application in which an SQL query, stored in a string variable, is sent to the database for execution. The model used is a non-deterministic finite automaton. The alphabet of the automaton is SQL keywords and operators, delimiters and place holders for input string values. After that, at runtime, all dynamically generated queries are monitored and checked for compliance with previously statically generated automaton model. Queries that violate the model are classified as illegal, prevented from executing on the database, and reported to the application developers and administrators.

Saner is a tool [13] that combines dynamic and static techniques to verify PHP programs. It supports language-based replacement by incorporating FSA tool [106] and if a sanitizer is found to be vulnerable, then a dynamic analysis is performed to check using a predefined set of dangerous test cases if sanitization operations could miss any of these test cases. This approach only supports bounded computation for loops, and it approximates variables updated in a loop as arbitrary strings if the computation does not converge within a given fixed bound.

11.2   String Constraint Solvers

There has been significant amount of work on string constraint solving in recent years [2, 44, 53, 54, 63, 71, 73, 93, 104, 135]. Constraint solving has received a lot of attention since it is used by symbolic execution tools which have become very popular. String constraint solvers are used specifically to deal with constraints that involve string variables.

Decidability problem for string constraints extracted from path conditions for programs using .NET string library has been studied in [21]. The .NET string library is modeled using a first order language called *the string library language*. It has been shown that the satisfiability of the string library language where the length of string variables is fixed and the replace operation is omitted is decidable. If the length is not fixed but replace is still omitted then the decidability problem is open. If replace is introduced then the satisfiability problem is undecidable for constraints with multiple variables. Hence, it is known that for some classes of string constraints it is not possible to have both a sound and complete constraint solver, and it is necessary to use approximations.

There has been three main research directions in string constraint solving: (1) Automata based solvers that map string constraints to automata [12, 40, 53, 54, 96, 130, 133, 134], (2) Bounded solvers that use bounded encodings such as bit-vectors [21, 63, 71, 93], and (3) String constraint solvers that focus on combination of theories [3, 15, 105, 112, 135]. Automata-based constraint solvers use different types of automata encodings for constraints, and they use approximations for constraints that are non-regular. Bit-vector based solvers support core string operations such as equality, membership, concatenation, and string length equations. Additional complex operations can be encoded using core string operations to some extent. Bounding the length of the strings allows bit-vector based solvers to handle length equations effectively. Some string constraint solvers are based on Satisfiability-Modulo-Theories (SMT) frameworks, and in addition to supporting core string operations, these solvers can also support some complex string operations such as replace operation and combination of string and numeric constraints. Bit-vector based solvers have limited support for mixed string constraints. Compared to the bit-vector based approaches, SMT solvers support several different theories and they are more expressive in terms of mixed constraints.

11.2.1   Automata-Based Solvers

Symbolic execution has been used to perform string analysis on Java programs where tracing path constraints and encoding the values of string variables are handled using automata [96]. Symbolic execution has also been used to find SQL injection vulnerabilities in .NET applications, where automata are used to represent string constraints and support string-based replacement (as opposed to language-based replacement) [40] Finite state transducers have been used to model constraints in PHP web applications for test input generation [121]. This approach is based on concolic execution [95], where results of a concrete execution are used to collect constraints on program execution. These constraints are then used to generate new test cases.

An automata-based decision procedure for solving equations over regular language variables using partial state space construction has been presented in [53, 54]. Since this work uses a single track automata encoding, it can only provide an approximation for solving equations over string variables. One potential solution is using multi-track automata to model relations among string variables [133, 134]. Rex [110, 111] uses symbolic automata to solve string constraints involving regular expressions.

Automata based string constraint solving techniques we discuss in Chaps. 5 and 7 use a symbolic automata representation, and represent the solution sets of constraints using automata [12, 130, 133, 134]. Using multi-track automata, relational constraints on multiple variables can also be represented.

11.2.2   Bounded Solvers

In [21] a string constraint solver that solves string constraint in two phases is presented. First, a string constraint is abstracted into an integer constraint by replacing each string variable with an unquantified integer variable. After solving the generated integer constraint with an SMT solver, results are used to fix (i.e., bound) the lengths of the strings. Then the original string constraint is solved after fixing the length of strings variables in it.

HAMPI [63] allows string constraints to be specified as a membership in a context free language or a regular language. Then a higher bound on string variables' lengths is specified which converts the constraint into a constraint on a finite (i.e., regular) language. The ability to specify constraints with context free languages is only a convenience feature which makes it much easier to specify constraints on variables that hold a context free language values such as SQL queries. Given an input constraint, it is normalized into a *core string constraint* where each constraint is of the form $x = R$ or $x \neq R$ where R is a regular expression. A simple algorithm is provided to convert a bounded CFG into a regular expression. Then core constraints are translated into a quantifier-free logic of bit-vectors constraints which are passed into a special constraint solver called STP. If there is a solution, HAMPI decodes the output bit-vectors into a string solution.

A string constraint solver called Kaluza [1] was built on top of the HAMPI. It uses the same approach of bounding the lengths of the execution paths (by bounding loops) and using a bounded string solver. Kaluza is used by KUDZU—a symbolic execution framework for JavaScript—to solve string constraints in JavaScript and generate new input that is used to explore more execution paths.

On one hand, bounded solvers are able to handle a larger set of string operations and predicates compared to automata-based string solvers. However, since they bound the length of strings they may miss some solutions that we can catch especially given how well our automata-based algorithms scale well with length.

Automata based string constraint solving techniques we discuss in earlier chapters do not bound the length of the strings, and instead use automata to encode solution sets of constraints which can be arbitrarily large, and even infinite.

11.2.3 Combination of Theories

An earlier research focus in analysis of string manipulation in programs was size analysis. Size analysis focuses on statically identifying all possible lengths of a string expression at a program point. This type of analysis can be used to identify and eliminate buffer overflow errors in programs for example [36, 43, 116]. There have been extensions of automata-based string analysis techniques that can handle both string and numeric constraints in programs by representing linear arithmetic constraints as automata as we discussed in Chap. 7 [132].

More recently, there has been a lot of work on SMT-based string constraint solvers [3, 15, 105, 135] that provide decision procedures for various fragments of string theories and which can be integrated with other decision procedures within the SMT framework. These approaches are not strictly restricted to bounded strings like the bounded string constraint solvers and can determine satisfiability of string constraints for unbounded domains in some cases.

11.2.4 Model Counting String Constraint Solvers

A model counting constraint solver does not only determines if a constraint is satisfiable, but it also determines the number of solutions to the constraint within a given bound. SMC is a model counter for string constraints [75]. SMC utilizes generating functions to count the number of strings that match to an unambiguous regular expression. In general, SMC generates a model-count range which consists of an upper bound and a lower bound. SMC cannot determine the precise model count for a regular expression constraint such as $x \in (a|b)^*|ab$, and it cannot propagate string values across logical connectives which reduces its precision.

In Chap. 7 we discuss an automata-based model counting approach for string constraints, which reduces the model counting problem to path counting [12]. This approach has been implemented as the model counting string constraint solver ABC discussed in Chap. 10. Automata based representation used in ABC allows precise model counting across logical connectives.

11.3 Bug and Vulnerability Detection in Web Applications

As we discussed in Chap. 1, one of the main motivations for string analysis is the prevalence of string manipulation code in web applications. There has been a significant amount of research on bug and vulnerability detection in web applications, and we provide a brief survey below. We first present work done on the client-side then move on to the server-side.

11.3.1 Client-Side Analysis

Static analysis of Javascript programs has been an active research topic [9, 9, 47–49, 58, 59, 62]. Static control [48], information [28] and taint [47] flow analyses have been used for Javascript programs to detect security vulnerabilities. For example, GATEKEEPER [46] uses static analysis to verify the enforcement of security policies written in Datalog on JavaScript widgets.

In general, though, pure static analyses for Javascript suffer from loss of precision which hinder their applicability in practice. The reason is, Javascript is a dynamic language and the dynamic features of the language are used heavily [90]. Hybrid analyses that combine static and dynamic analyses, or combination of string analysis with string analysis may improve the precision and reduce the rate of false alarms. For example, since objects and arrays in Javascript are maps from strings to strings, string analysis maybe useful during static analysis.

In addition to static analysis, dynamic analysis techniques have been used in [1, 6, 35, 64, 94] to extract and/or analyze Javascript code. FLAX [94] uses dynamic analysis techniques to discover client side validation vulnerabilities. The authors use dynamic taint analysis to extract validation code related to a certain sink and then use random fuzzing to test this sink.

In [1] authors developed a symbolic execution framework for JavaScript. At the core of their framework there is a string constraint solver called KUDZU that is built on top of the bounded string solver HAMPI [63]. Static string analysis can be combined with dynamic extraction techniques in order to apply it to JavaScript code which is hard to analyze using only static techniques [6].

11.3.2 Server-Side Analysis

There has been many static vulnerability detection techniques that have been developed for PHP and Java web applications. Many of these techniques such as [26, 78, 103, 118–120] are based exclusively on static string analysis and we discuss them in more detail in the next section. Pixy [60] uses different static analysis techniques to build dependency graphs that represent the data flow from sources to sinks in a PHP web application. Then, it uses taint analysis to detect if there are vulnerabilities in web applications. The STRANGER [127] and SEMREP [5] tools discussed in Chap. 10 use Pixy as a front end, and improve the precision of vulnerability analysis by using string analysis techniques [5, 7, 126, 128].

In [124], the problem of statically detecting SQL injection vulnerabilities in PHP scripts is addressed using a three-tier approach. Information is computed bottom-up for the intra-block, intra-procedural, and inter-procedural scope. As a result, the analysis is flow-sensitive and inter-procedural. Traditional data flow analysis is used to determine whether unchecked user inputs can reach security-sensitive functions (called sinks) without being properly checked. However, any information about the possible strings that a variable might hold is not computed. Thus, some types of vulnerabilities might be missed. RIPS [31] uses the same technique with extensive modeling for PHP built-in functions and libraries and extends this approach to other types of vulnerabilities such cross-site scripting (XSS) and malicious file execution (MFE). Phantm [67] runs the PHP web application to solve include statements and configuration values, and then uses the dynamically collected data to improve precision of static analysis. Apollo [10] uses dynamic symbolic execution to generate test cases for the web application. It applies some techniques to minimize the conditions on the inputs that cause a failure to provide better error reporting. Dynamic symbolic execution along with grammar based string analysis has also been used to generate test inputs for PHP web applications [121]. Saner [13] mixes string-based static and dynamic techniques to discover vulnerabilities.

11.4 Differential Analysis and Repair

Analyzing differences between code segments can be useful in many contexts. As we have seen with client- and server-side code in web applications, sometimes one can use one part of code as a specification for another part of code. Another scenario would be to provide reference validation or sanitization code that developers can use in implementing their own validation and sanitization code. Yet another scenario would be to determine the affects of changes between different versions of code. Differential analysis and repair techniques automatically analyze code segments with respect to each other in order to identify bugs and vulnerabilities or to repair them.

11.4.1 Differential Analysis

Differential analysis techniques [68, 70, 87] typically stop after finding differences between different pieces of code without trying to repair it. In [87], differential symbolic execution is used to find differences between original and refactored code by summarizing procedures into symbolic constraints and then comparing different summaries using an SMT solver. SYMDIFF [68] computes the difference between two different functions in a language agnostic way by reducing both functions to Boggie [14] intermediate language then finds semantic differences using the Z3 SMT solver [76].

There are several specialized differential analysis techniques that focus on web applications. In NoTamper [19] the authors analyze client-side script code using dynamic symbolic execution to generate test cases that are subsequently used as inputs to the server side of the application. Since the approach relies on dynamic (black-box) testing, it can suffer from limited code coverage. In a recent follow up paper [20], a new tool called WAPTEC is presented, which uses symbolic execution of the server code to guide the test case generation process and expand coverage. MiTV [101] uses dynamic symbolic execution engine Pex [77] to test the correctness of user input validation functions for .NET web applications. These functions are first classified according to the type of input they validate. Then each validation function is tested by comparing it to a subset of the functions under the same class.

Differential string analysis techniques we discussed in Chap. 9 involve extraction of client- and server-side validation and sanitization routines and their analysis using automata-based string analysis techniques [5, 7].

11.4.2   Differential Repair

Differential program repair [8, 81, 92, 97, 98, 122, 123] became an active topic recently. In [122, 123], detected bugs are repaired based on manually written test suites using genetic algorithms. The abstract syntax tree (AST) of the program is randomly mutated multiple times by deleting, swapping and/or copying subtrees related to the execution path taken by the test suite. Mutation is done until a mutated version passes the original test suite. Correctness of repairs generated by this approach depends on the effectiveness of the test suite.

In [97, 98], access control problems in PHP programs are detected by comparing a possibly buggy AST with one that is considered to be correct and then patch the difference by inserting statements from the latter into the former. In contrast to these syntactic repair approaches, the differential repair techniques we discuss in Chap. 9 rely on semantic analysis of the code [5, 7].

In [81, 92], test suites are used to find bugs then symbolic execution is used to find constraints on variables that result in such bugs. Using the solution to the negation of these constraints, a patch is synthesized for the program such that it passes all test suites.

In [74] a set of sanitizers are automatically placed in a sanitizer free program based on a user defined policy and a flow graph. The sanitizers are placed in the flow graph such that they satisfy the specified policy and at the same time avoid idempotency problems. The techniques we discussed in Chap. 9 [5, 7] takes the existing sanitization code into account and places the repair/patch before the original code, instead of changing the code. This avoids interference with the original sanitization code that may have side-effects.

Chapter 12
Conclusions

String manipulation is an increasingly important part of modern software development. And, it can be error prone due to complexity of string expressions. Many software bugs are caused by errors in manipulation of string values in programs. Moreover, string manipulation errors can be very costly. One of the leading causes of software security vulnerabilities is string manipulation errors. We argue that automated analysis of string expressions in programs is one of the critical problems in software dependability.

In this monograph we presented automata-based symbolic string analysis techniques for finding and eliminating bugs and security vulnerabilities in string manipulating programs.

After presenting an overview of common scenarios for using string manipulation in programs, we gave examples of common errors and vulnerabilities that are due to string manipulation. We showed that precise automated analysis of strings is not possible in general even for very basic string operations such as concatenation. We also provided an overview of more complex string operations that are commonly used in programming.

We gave an overview of state space exploration techniques, including explicit state, symbolic, forward and backward analyses and fixpoint computations. We argued that automata can be used as a symbolic representation for string analysis, and we presented a symbolic encoding of automata. We discussed how pre- and post-condition of common string manipulation operations can be computed using automata, and how they can be used in symbolic reachability analysis to compute possible values that string expressions can take during program execution.

We discussed two extensions to basic symbolic reachability analysis for string expressions using automata: (1) Relational string analysis using multi-track automata which enables us to keep track of relations among string variables during reachability analysis; and (2) Automated abstraction and approximation techniques which enable us to limit the blow-up experienced during static string analysis and accelerate fixpoint computations in order to achieve convergence.

© Springer International Publishing AG 2017
T. Bultan et al., *String Analysis for Software Verification and Security*,
https://doi.org/10.1007/978-3-319-68670-7_12

Next, we discussed automata-based string constraint solving techniques. In the last decade, symbolic execution has become the dominant technology for verification of software. However, ability of symbolic execution tools in analyzing string manipulating code has been very limited. Novel string constraint solving techniques can improve applicability of symbolic execution to string manipulating code.

We also discussed model counting for string constraints. Model counting is a crucial problem for quantitative program analysis. By mapping string constraints to automata, we can reduce model counting problem to path counting and use existing techniques from algebraic graph theory for model counting for string constraints.

We discussed how automata-based string analysis techniques can be used for vulnerability detection in web applications. Furthermore, we showed that automata-based string analysis techniques can be used to synthesize sanitizers that can be used to repair vulnerabilities in existing code.

We extended the discussion on vulnerability detection and repair to differential analysis where the idea is to detect vulnerabilities by detecting inconsistencies between different pieces of code, and repair the vulnerabilities by eliminating identified inconsistencies.

We gave an overview of a set of tools that implement the automata-based string analysis techniques we discussed. Furthermore, we provided a survey of related results from the research literature.

We believe that string analysis will continue to be an important problem in software dependability. Scalable string analysis techniques are sorely needed and we hope that the topics we discussed in this monograph would inspire other researchers to make further contributions in this area.

References

1. Prateek Saxena, Devdatta Akhawe, Steve Hanna, Feng Mao, Stephen McCamant, and Dawn Song. A symbolic execution framework for javascript. In *Proceedings of the IEEE Symposium on Security and Privacy,* pages 513–528, 2010.
2. Parosh Aziz Abdulla, Mohamed Faouzi Atig, Yu-Fang Chen, Lukás Holík, Ahmed Rezine, Philipp Rümmer, and Jari Stenman. String constraints for verification. In *Proceedings of the 26th International Conference on Computer Aided Verification (CAV),* pages 150–166, 2014.
3. Parosh Aziz Abdulla, Mohamed Faouzi Atig, Yu-Fang Chen, Lukáš Holík, Ahmed Rezine, Philipp Rümmer, and Jari Stenman. *Computer Aided Verification: 27th International Conference, CAV 2015, San Francisco, CA, USA, July 18–24, 2015, Proceedings, Part I,* chapter Norn: An SMT Solver for String Constraints, pages 462–469. Springer International Publishing, Cham, 2015.
4. Muath Alkhalaf. *Automatic Detection and Repair of Input Validation and Sanitization Bugs.* Dissertation, University of California Santa Barbara, 2014.
5. Muath Alkhalaf, Abdulbaki Aydin, and Tevfik Bultan. Semantic differential repair for input validation and sanitization. In *Proceedings of the 2014 International Symposium on Software Testing and Analysis (ISSTA 2014),* 2014.
6. Muath Alkhalaf, Tevfik Bultan, and Jose L. Gallegos. Verifying client-side input validation functions using string analysis. In *Proceedings of the 2012 International Conference on Software Engineering,* pages 947–957, 2012.
7. Muath Alkhalaf, Shauvik Roy Choudhary, Mattia Fazzini, Tevfik Bultan, Alessandro Orso, and Christopher Kruegel. Viewpoints: differential string analysis for discovering client- and server-side input validation inconsistencies. In *Proceedings of the 2012 International Symposium on Software Testing and Analysis (ISSTA),* pages 56–66, 2012.
8. Jesper Andersen and Julia L. Lawall. Generic patch inference. In *Proceedings of the 2008 23rd IEEE/ACM International Conference on Automated Software Engineering,* ASE '08, pages 337–346, Washington, DC, USA, 2008. IEEE Computer Society.
9. Christopher Anderson, Paola Giannini, and Sophia Drossopoulou. Towards type inference for javascript. In *ECOOP 2005-Object-Oriented Programming,* pages 428–452. Springer, 2005.
10. Shay Artzi, Adam Kiezun, Julian Dolby, Frank Tip, Daniel Dig, Amit Paradkar, and Michael D Ernst. Finding bugs in web applications using dynamic test generation and explicit-state model checking. *Software Engineering, IEEE Transactions on,* 36(4):474–494, 2010.

© Springer International Publishing AG 2017

T. Bultan et al., *String Analysis for Software Verification and Security,*
https://doi.org/10.1007/978-3-319-68670-7

11. Abdulbaki Aydin. *Automata-based Model Counting String Constraint Solver for Vulnerability Analysis*. Dissertation, University of California Santa Barbara, 2017.
12. Abdulbaki Aydin, Lucas Bang, and Tevfik Bultan. *Computer Aided Verification: 27th International Conference, CAV 2015, San Francisco, CA, USA, July 18–24, 2015, Proceedings, Part I*, chapter Automata-Based Model Counting for String Constraints, pages 255–272. Springer International Publishing, Cham, 2015.
13. Davide Balzarotti, Marco Cova, Vika Felmetsger, Nenad Jovanovic, Engin Kirda, Christopher Kruegel, and Giovanni Vigna. Saner: Composing static and dynamic analysis to validate sanitization in web applications. In *Proceedings of the 2008 IEEE Symposium on Security and Privacy*, SP '08, pages 387–401, Washington, DC, USA, 2008. IEEE Computer Society.
14. Mike Barnett, Bor-Yuh Evan Chang, Robert DeLine, Bart Jacobs, and K. Rustan M. Leino. Boogie: A modular reusable verifier for object-oriented programs. In *Proceedings of the 4th International Conference on Formal Methods for Components and Objects*, FMCO'05, pages 364–387, Berlin, Heidelberg, 2006. Springer-Verlag.
15. Clark Barrett, Christopher L. Conway, Morgan Deters, Liana Hadarean, Dejan Jovanović, Tim King, Andrew Reynolds, and Cesare Tinelli. *Computer Aided Verification: 23rd International Conference, CAV 2011, Snowbird, UT, USA, July 14–20, 2011. Proceedings*, chapter CVC4, pages 171–177. Springer Berlin Heidelberg, Berlin, Heidelberg, 2011.
16. Constantinos Bartzis and Tevfik Bultan. Efficient symbolic representations for arithmetic constraints in verification. *International Journal of Foundations of Computer Science (IJFCS)*, 14(4):605–624, August 2003.
17. Constantinos Bartzis and Tevfik Bultan. Widening arithmetic automata. In R. Alur and D. Peled, editors, *Proceedings of the 16th International Conference on Computer Aided Verification (CAV 2004)*, volume 3114 of *Lecture Notes in Computer Science*, pages 321–333. Springer-Verlag, July 2004.
18. Norman Biggs. *Algebraic Graph Theory*. Cambridge Mathematical Library. Cambridge University Press, 1993.
19. Prithvi Bisht, Timothy Hinrichs, Nazari Skrupsky, Radoslaw Bobrowicz, and V. N. Venkatakrishnan. Notamper: automatic blackbox detection of parameter tampering opportunities in web applications. In *Proceedings of the 17th ACM conference on Computer and communications security*, CCS '10, pages 607–618, New York, NY, USA, 2010. ACM.
20. Prithvi Bisht, Timothy Hinrichs, Nazari Skrupsky, and V. N. Venkatakrishnan. Waptec: Whitebox analysis of web applications for parameter tampering exploit construction. In *Proceedings of the 18th ACM Conference on Computer and Communications Security*, CCS '11, pages 575–586, New York, NY, USA, 2011. ACM.
21. Nikolaj Bjørner, Nikolai Tillmann, and Andrei Voronkov. Path feasibility analysis for string-manipulating programs. In *Proceedings of the 15th International Conference on Tools and Algorithms for the Construction and Analysis of Systems: Held As Part of the Joint European Conferences on Theory and Practice of Software, ETAPS 2009,*, TACAS '09, pages 307–321, Berlin, Heidelberg, 2009. Springer-Verlag.
22. Eric Bodden, Andreas Sewe, Jan Sinschek, Hela Oueslati, and Mira Mezini. Taming reflection: Aiding static analysis in the presence of reflection and custom class loaders. In *Proceedings of the 33rd International Conference on Software Engineering*, ICSE '11, pages 241–250, New York, NY, USA, 2011. ACM.
23. Ahmed Bouajjani, Peter Habermehl, and Tomáš Vojnar. Abstract regular model checking. In Rajeev Alur and Doron A. Peled, editors, *Computer Aided Verification: 16th International Conference, CAV 2004, Boston, MA, USA, July 13–17, 2004. Proceedings*, pages 372–386, Berlin, Heidelberg, 2004. Springer Berlin Heidelberg.
24. BRICS. The MONA project. http://www.brics.dk/mona/.
25. Tae-Hyoung Choi, Oukseh Lee, Hyunha Kim, and Kyung-Goo Doh. A practical string analyzer by the widening approach. In *APLAS*, pages 374–388, 2006.
26. Aske Simon Christensen, Anders Møller, and Michael I. Schwartzbach. Precise analysis of string expressions. In *Proc. 10th International Static Analysis Symposium, SAS '03*, volume 2694 of *LNCS*, pages 1–18. Springer-Verlag, June 2003.

27. Mihai Christodorescu, Nicholas Kidd, and Wen-Han Goh. String analysis for x86 binaries. In *Proceedings of the 6th ACM SIGPLAN-SIGSOFT Workshop on Program Analysis for Software Tools and Engineering (PASTE 2005)*. ACM Press, September 2005.

28. Ravi Chugh, Jeffrey A Meister, Ranjit Jhala, and Sorin Lerner. Staged information flow for javascript. In *ACM Sigplan Notices*, volume 44, pages 50–62. ACM, 2009.

29. Thomas H. Cormen, Charles E. Leiserson, and Ronald L. Rivest. *Introduction to Algorithms*. MIT Press, 1990.

30. CVE. Common Vulnerabilities and Exposures. http://www.cve.mitre.org.

31. Johannes Dahse and Thorsten Holz. Simulation of built-in php features for precise static code analysis. In *Proceedings of Network and Distributed System Security (NDSS'14) Symposium*, 2014.

32. Loris D'Antoni and Margus Veanes. Equivalence of extended symbolic finite transducers. In *Computer Aided Verification*, pages 624–639. Springer, 2013.

33. Loris D'Antoni and Margus Veanes. Minimization of symbolic automata. In *Proceedings of the 41st annual ACM SIGPLAN-SIGACT symposium on Principles of programming languages*, pages 541–554. ACM, 2014.

34. Loris D'antoni and Margus Veanes. Extended symbolic finite automata and transducers. *Form. Methods Syst. Des.*, 47(1):93–119, August 2015.

35. Mohan Dhawan and Vinod Ganapathy. Analyzing information flow in javascript-based browser extensions. In *Computer Security Applications Conference, 2009. ACSAC'09. Annual*, pages 382–391. IEEE, 2009.

36. Nurit Dor, Michael Rodeh, and Mooly Sagiv. Cssv: towards a realistic tool for statically detecting all buffer overflows in c. *SIGPLAN Not.*, 38(5):155–167, 2003.

37. DroidBench. Droidbench benchmarks. https://github.com/secure-software-engineering/DroidBench.

38. Loris D'Antoni and Margus Veanes. Static analysis of string encoders and decoders. In *Proceedings of the 14th International Conference on Verification, Model Checking, and Abstract Interpretation (VMCAI)*, pages 209–228, 2013.

39. Philippe Flajolet and Robert Sedgewick. *Analytic Combinatorics*. Cambridge University Press, New York, NY, USA, 1 edition, 2009.

40. Xiang Fu, Xin Lu, Boris Peltsverger, Shijun Chen, Kai Qian, and Lixin Tao. A static analysis framework for detecting sql injection vulnerabilities. In *COMPSAC*, pages 87–96, 2007.

41. Masahiro Fujita, Patrick C. McGeer, and Jerry Chih-Yuan Yang. Multi-terminal binary decision diagrams: An efficient data structure for matrix representation. *Formal Methods in System Design*, 10(2/3):149–169, 1997.

42. William G. J. Halfond, Jeremy Viegas, and Alessandro Orso. A classification of sql injection attacks and countermeasures. In *Proceedings of the International Symposium on Secure Software Engineering*, 2006.

43. Vinod Ganapathy, Somesh Jha, David Chandler, David Melski, and David Vitek. Buffer overrun detection using linear programming and static analysis. In *Proceedings of the 10th ACM Conference on Computer and Communications Security*, pages 345–354, 2003.

44. Vijay Ganesh, Mia Minnes, Armando Solar-Lezama, and Martin C. Rinard. Word equations with length constraints: What's decidable? In *Proceedings of the 8th International Haifa Verification Conference (HVC)*, pages 209–226, 2012.

45. Dale Gerdemann and Gertjan van Noord. Transducers from rewrite rules with backreferences. In *Proceedings of the 9th Conference of the European Chapter of the Association for Computational Linguistics*, pages 126–133, 1999.

46. Salvatore Guarnieri and Benjamin Livshits. Gatekeeper: mostly static enforcement of security and reliability policies for javascript code. In *Proceedings of the 18th conference on USENIX security symposium*, SSYM'09, pages 151–168, Berkeley, CA, USA, 2009. USENIX Association.

47. Salvatore Guarnieri, Marco Pistoia, Omer Tripp, Julian Dolby, Stephen Teilhet, and Ryan Berg. Saving the world wide web from vulnerable javascript. In *Proceedings of the 2011 International Symposium on Software Testing and Analysis*, pages 177–187. ACM, 2011.

48. Arjun Guha, Shriram Krishnamurthi, and Trevor Jim. Static analysis for ajax intrusion detection. In *Proceedings of the International World Wide Web Conference*. Citeseer, 2009.

49. Arjun Guha, Claudiu Saftoiu, and Shriram Krishnamurthi. The essence of javascript. In *ECOOP 2010–Object-Oriented Programming*, pages 126–150. Springer, 2010.

50. William G. J. Halfond and Alessandro Orso. Amnesia: analysis and monitoring for neutralizing sql-injection attacks. In *ASE '05: Proceedings of the 20th IEEE/ACM international Conference on Automated software engineering*, pages 174–183, New York, NY, USA, 2005. ACM.

51. Jesper G. Henriksen, Jakob Jensen, Michael Jørgensen, Nils Klarlund, Robert Paige, Theis Rauhe, and Anders Sandholm. Mona: Monadic second-order logic in practice. In E. Brinksma, W. R. Cleaveland, K. G. Larsen, T. Margaria, and B. Steffen, editors, *Tools and Algorithms for the Construction and Analysis of Systems: First International Workshop, TACAS '95 Aarhus, Denmark, May 19–20, 1995 Selected Papers*, pages 89–110, Berlin, Heidelberg, 1995. Springer Berlin Heidelberg.

52. Pieter Hooimeijer, Ben Livshits, David Molnar, Prateek Saxena, and Margus Veanes. Fast and Precise Sanitizer Analysis with Bek. In *Usenix Security Symposium*, 2011.

53. Pieter Hooimeijer and Westley Weimer. A decision procedure for subset constraints over regular languages. In *Proceedings of the ACM SIGPLAN Conference on Programming Language Design and Implementation (PLDI)*, pages 188–198, 2009.

54. Pieter Hooimeijer and Westley Weimer. Solving string constraints lazily. In *Proceedings of the 25th IEEE/ACM International Conference on Automated Software Engineering (ASE)*, pages 377–386, 2010.

55. Pieter Hooimeijer and Westley Weimer. Strsolve: solving string constraints lazily. *Automated Software Engineering*, 19(4):531–559, 2012.

56. John E. Hopcroft, Rajeev Motwani, and Jeffrey D. Ullman. *Introduction to Automata Theory, Languages, and Computation (3rd Edition)*. Addison-Wesley Longman Publishing Co., Inc., Boston, MA, USA, 2006.

57. Simon Holm Jensen, Peter A. Jonsson, and Anders Møller. Remedying the eval that men do. In *Proceedings of the 2012 International Symposium on Software Testing and Analysis (ISSTA)*, pages 34–44, 2012.

58. Simon Holm Jensen, Magnus Madsen, and Anders Møller. Modeling the html dom and browser api in static analysis of javascript web applications. In *Proceedings of the 19th ACM SIGSOFT symposium and the 13th European conference on Foundations of software engineering*, pages 59–69. ACM, 2011.

59. Simon Holm Jensen, Anders Møller, and Peter Thiemann. Type analysis for javascript. In *Static Analysis*, pages 238–255. Springer, 2009.

60. Nenad Jovanovic, Christopher Krügel, and Engin Kirda. Pixy: A static analysis tool for detecting web application vulnerabilities. In *S&P*, pages 258–263, 2006.

61. Lauri Karttunen. The replace operator. In *Proceedings of the 33rd annual meeting on Association for Computational Linguistics*, pages 16–23, 1995.

62. Vineeth Kashyap, John Sarracino, John Wagner, Ben Wiedermann, and Ben Hardekopf. Type refinement for static analysis of javascript. In *Proceedings of the 9th symposium on Dynamic languages*, pages 17–26. ACM, 2013.

63. Adam Kiezun, Vijay Ganesh, Philip J. Guo, Pieter Hooimeijer, and Michael D. Ernst. Hampi: a solver for string constraints. In *Proceedings of the 18th International Symposium on Software Testing and Analysis (ISSTA)*, pages 105–116, 2009.

64. Haruka Kikuchi, Dachuan Yu, Ajay Chander, Hiroshi Inamura, and Igor Serikov. Javascript instrumentation in practice. In *Programming Languages and Systems*, pages 326–341. Springer, 2008.

65. James C. King. Symbolic execution and program testing. *Commun. ACM*, 19(7):385–394, July 1976.

66. Christian Kirkegaard, Anders Møller, and Michael I. Schwartzbach. Static analysis of xml transformations in java. *IEEE Transactions on Software Engineering*, 30(3), March 2004.

67. Etienne Kneuss, Philippe Suter, and Viktor Kuncak. Phantm: Php analyzer for type mismatch. In *Proceedings of the Eighteenth ACM SIGSOFT International Symposium on Foundations of Software Engineering*, FSE '10, pages 373–374, New York, NY, USA, 2010. ACM.

68. Shuvendu K. Lahiri, Chris Hawblitzel, Ming Kawaguchi, and Henrique Rebêlo. Symdiff: A language-agnostic semantic diff tool for imperative programs. In *Proceedings of the 24th International Conference on Computer Aided Verification (CAV)*, pages 712–717, 2012.

69. Shuvendu K. Lahiri, Kenneth L. McMillan, Rahul Sharma, and Chris Hawblitzel. Differential assertion checking. In *Proceedings of the 2013 9th Joint Meeting on Foundations of Software Engineering (ESEC/FSE)*, pages 345–355, 2013.

70. Shuvendu K. Lahiri, Kapil Vaswani, and C A. R. Hoare. Differential static analysis: Opportunities, applications, and challenges. In *Proceedings of the FSE/SDP Workshop on Future of Software Engineering Research*, pages 201–204, 2010.

71. Guodong Li and Indradeep Ghosh. PASS: string solving with parameterized array and interval automaton. In *Proceedings of the 9th International Haifa Verification Conference (HVC)*, pages 15–31, 2013.

72. Li Li, Tegawendé F Bissyandé, Damien Octeau, and Jacques Klein. Droidra: taming reflection to support whole-program analysis of android apps. In *Proceedings of the 25th International Symposium on Software Testing and Analysis*, pages 318–329. ACM, 2016.

73. Tianyi Liang, Andrew Reynolds, Cesare Tinelli, Clark Barrett, and Morgan Deters. A DPLL(T) theory solver for a theory of strings and regular expressions. In *Proceedings of the 26th International Conference on Computer Aided Verification (CAV)*, pages 646–662, 2014.

74. Benjamin Livshits and Stephen Chong. Towards fully automatic placement of security sanitizers and declassifiers. In *Proceedings of the 40th Annual ACM SIGPLAN-SIGACT Symposium on Principles of Programming Languages*, POPL '13, pages 385–398, New York, NY, USA, 2013. ACM.

75. Loi Luu, Shweta Shinde, Prateek Saxena, and Brian Demsky. A model counter for constraints over unbounded strings. In *Proceedings of the ACM SIGPLAN Conference on Programming Language Design and Implementation (PLDI)*, page 57, 2014.

76. Microsoft Inc. Z3 SMT Solver. http://z3.codeplex.com.

77. Microsoft Research. Pex. http://research.microsoft.com/en-us/projects/pex/.

78. Yasuhiko Minamide. Static approximation of dynamically generated web pages. In *Proceedings of the 14th International World Wide Web Conference (WWW)*, pages 432–441, 2005.

79. Marvin L. Minsky. Recursive unsolvability of Post's problem of Tag and other topics in the theory of Turing machines. In *Ann. of Math (74)*, pages 437–455, 1961.

80. Mehryar Mohri and Richard Sproat. An efficient compiler for weighted rewrite rules. In *Proceedings of the 34th annual meeting on Association for Computational Linguistics*, pages 231–238. Association for Computational Linguistics, 1996.

81. Hoang Duong Thien Nguyen, Dawei Qi, Abhik Roychoudhury, and Satish Chandra. Semfix: Program repair via semantic analysis. In *Proceedings of the 2013 International Conference on Software Engineering*, ICSE '13, pages 772–781, Piscataway, NJ, USA, 2013. IEEE Press.

82. Hung Viet Nguyen, Christian Kästner, and Tien N. Nguyen. Building call graphs for embedded client-side code in dynamic web applications. In *Proceedings of the 22nd ACM SIGSOFT International Symposium on Foundations of Software Engineering (FSE-22)*, pages 518–529, 2014.

83. Hung Viet Nguyen, Christian Kästner, and Tien N. Nguyen. Varis: IDE support for embedded client code in PHP web applications. In *Proceedings of the 37th IEEE/ACM International Conference on Software Engineering (ICSE)*, pages 693–696, 2015.

84. OWASP. Top 10 2007. https://www.owasp.org/index.php/Top_10_2007.

85. OWASP. Top 10 2010. https://www.owasp.org/index.php/Top_10_2010-Main.

86. OWASP. Top 10 2013. https://www.owasp.org/index.php/Top_10_2013-T10.

87. Suzette J. Person. *Differential Symbolic Execution*. PhD thesis, Lincoln, NB, USA, 2009. AAI3365729.

88. Bala Ravikumar and Gerry Eisman. Weak minimization of DFA - an algorithm and applications. *Theor. Comput. Sci.*, 328(1–2):113–133, 2004.
89. Gideon Redelinghuys, Willem Visser, and Jaco Geldenhuys. Symbolic execution of programs with strings. In *Proceedings of the South African Institute for Computer Scientists and Information Technologists Conference*, SAICSIT '12, pages 139–148, New York, NY, USA, 2012. ACM.
90. Gregor Richards, Sylvain Lebresne, Brian Burg, and Jan Vitek. An analysis of the dynamic behavior of javascript programs. In *Proceedings of the 2010 ACM SIGPLAN conference on Programming language design and implementation*, PLDI '10, pages 1–12, New York, NY, USA, 2010. ACM.
91. Yuto Sakuma, Yasuhiko Minamide, and Andrei Voronkov. Translating regular expression matching into transducers. *J. Applied Logic*, 10(1):32–51, 2012.
92. Hesam Samimi, Max Schäfer, Shay Artzi, Todd Millstein, Frank Tip, and Laurie Hendren. Automated repair of html generation errors in php applications using string constraint solving. In *Proceedings of the 2012 International Conference on Software Engineering*, ICSE 2012, pages 277–287, Piscataway, NJ, USA, 2012. IEEE Press.
93. Prateek Saxena, Devdatta Akhawe, Steve Hanna, Feng Mao, Stephen McCamant, and Dawn Song. A symbolic execution framework for javascript. In *Proceedings of the 31st IEEE Symposium on Security and Privacy*, 2010.
94. Prateek Saxena, Steve Hanna, Pongsin Poosankam, and Dawn Song. Flax: Systematic discovery of client-side validation vulnerabilities in rich web applications. In *Proceedings of the Network and Distributed System Security Symposium (NDSS)*, 2010.
95. Koushik Sen, Darko Marinov, and Gul Agha. Cute: a concolic unit testing engine for c. In *Proceedings of the 10th European Software Engineering Conference held jointly with 13th ACM SIGSOFT International Symposium on Foundations of Software Engineering (ESEC/FSE 05)*, pages 263–272, 2005.
96. Daryl Shannon, Sukant Hajra, Alison Lee, Daiqian Zhan, and Sarfraz Khurshid. Abstracting symbolic execution with string analysis. In *TAICPART-MUTATION*, pages 13–22, 2007.
97. Sooel Son, Kathryn S. McKinley, and Vitaly Shmatikov. Rolecast: Finding missing security checks when you do not know what checks are. In *Proceedings of the 2011 ACM International Conference on Object Oriented Programming Systems Languages and Applications*, OOPSLA '11, pages 1069–1084, New York, NY, USA, 2011. ACM.
98. Sooel Son, Kathryn S. McKinley, and Vitaly Shmatikov. Fix me up: Repairing access-control bugs in web applications. In *NDSS*, 2013.
99. Richard P. Stanley. *Enumerative Combinatorics: Volume 1*. Cambridge University Press, New York, NY, USA, 2nd edition, 2011.
100. Zhendong Su and Gary Wassermann. The essence of command injection attacks in web applications. In *Conference Record of the 33rd ACM SIGPLAN-SIGACT Symposium on Principles of Programming Languages*, POPL '06, pages 372–382, New York, NY, USA, 2006. ACM.
101. Kunal Taneja, Nuo Li, Madhuri R. Marri, Tao Xie, and Nikolai Tillmann. Mitv: multiple-implementation testing of user-input validators for web applications. In *ASE*, pages 131–134, 2010.
102. Alfred Tarski. A lattice-theoretical fixpoint theorem and its applications. *Pacific Journal of Mathematics*, 5:285–309, 1955.
103. Takaaki Tateishi, Marco Pistoia, and Omer Tripp. Path- and index-sensitive string analysis based on monadic second-order logic. In *Proceedings of the 2011 International Symposium on Software Testing and Analysis*, ISSTA '11, pages 166–176, New York, NY, USA, 2011. ACM.
104. Minh-Thai Trinh, Duc-Hiep Chu, and Joxan Jaffar. S3: A symbolic string solver for vulnerability detection in web applications. In *Proceedings of the ACM SIGSAC Conference on Computer and Communications Security (CCS)*, pages 1232–1243, 2014.

105. Minh-Thai Trinh, Duc-Hiep Chu, and Joxan Jaffar. S3: A symbolic string solver for vulnerability detection in web applications. In *Proceedings of the 2014 ACM SIGSAC Conference on Computer and Communications Security*, CCS '14, pages 1232–1243, New York, NY, USA, 2014. ACM.

106. Gertjan van Noord. FSA utilities toolbox. http://odur.let.rug.nl/~vannoord/Fsa/.

107. Gertjan van Noord and Dale Gerdemann. An extendible regular expression compiler for finite-state approaches in natural language processing. In *Proc. of the 4th International Workshop on Implementing Automata (WIA)*, pages 122–139. Springer-Verlag, July 1999.

108. Margus Veanes. Symbolic string transformations with regular lookahead and rollback. In *Proceedings of the 9th Ershov Informatics Conference (PSI'14)*. Springer, 2014.

109. Margus Veanes and Nikolaj Bjørner. Symbolic automata: The toolkit. In *TACAS*, pages 472–477, 2012.

110. Margus Veanes, Nikolaj Bjørner, and Leonardo De Moura. Symbolic automata constraint solving. In *Logic for Programming, Artificial Intelligence, and Reasoning*, pages 640–654. Springer, 2010.

111. Margus Veanes, Peli De Halleux, and Nikolai Tillmann. Rex: Symbolic regular expression explorer. In *Software Testing, Verification and Validation (ICST), 2010 Third International Conference on*, pages 498–507. IEEE, 2010.

112. Margus Veanes, Peli de Halleux, and Nikolai Tillmann. Rex: Symbolic regular expression explorer. In *Proceedings of the 2010 Third International Conference on Software Testing, Verification and Validation*, ICST '10, pages 498–507, Washington, DC, USA, 2010. IEEE Computer Society.

113. Margus Veanes, Pieter Hooimeijer, Benjamin Livshits, David Molnar, and Nikolaj Bjorner. Symbolic finite state transducers: algorithms and applications. In *Proceedings of the 39th annual ACM SIGPLAN-SIGACT symposium on Principles of programming languages*, POPL '12, pages 137–150, New York, NY, USA, 2012. ACM.

114. Margus Veanes, Todd Mytkowicz, David Molnar, and Benjamin Livshits. Data-parallel string-manipulating programs. In *Proceedings of the 42Nd Annual ACM SIGPLAN-SIGACT Symposium on Principles of Programming Languages*, POPL '15, pages 139–152, New York, NY, USA, 2015. ACM.

115. David Wagner, Jeffrey S. Foster, Eric A. Brewer, and Alexander Aiken. A first step towards automated detection of buffer overrun vulnerabilities. In *Proc. of the Network and Distributed System Security Symposium*, pages 3–17, 2000.

116. David Wagner, Jeffrey S. Foster, Eric A. Brewer, and Alexander Aiken. A first step towards automated detection of buffer overrun vulnerabilities. In *In Network and Distributed System Security Symposium*, pages 3–17, 2000.

117. Hung-En Wang, Tzung-Lin Tsai, Chun-Han Lin, Fang Yu, and Jie-Hong R. Jiang. String analysis via automata manipulation with logic circuit representation. In *Computer Aided Verification - 28th International Conference, CAV 2016, Toronto, ON, Canada, July 17–23, 2016, Proceedings, Part I*, pages 241–260, 2016.

118. Gary Wassermann, Carl Gould, Zhendong Su, and Premkumar Devanbu. Static checking of dynamically generated queries in database applications. volume 16, New York, NY, USA, September 2007. ACM.

119. Gary Wassermann and Zhendong Su. Sound and precise analysis of web applications for injection vulnerabilities. In *Proceedings of the ACM SIGPLAN 2007 Conference on Programming Language Design and Implementation (PLDI)*, pages 32–41, 2007.

120. Gary Wassermann and Zhendong Su. Static detection of cross-site scripting vulnerabilities. In *Proceedings of the 30th International Conference on Software Engineering*, ICSE '08, pages 171–180, New York, NY, USA, 2008. ACM.

121. Gary Wassermann, Dachuan Yu, Ajay Chander, Dinakar Dhurjati, Hiroshi Inamura, and Zhendong Su. Dynamic test input generation for web applications. In *Proceedings of the ACM/SIGSOFT International Symposium on Software Testing and Analysis (ISSTA 2008)*, pages 249–260, 2008.

122. Westley Weimer, Stephanie Forrest, Claire Le Goues, and ThanhVu Nguyen. Automatic program repair with evolutionary computation. *Commun. ACM*, 53(5):109–116, May 2010.

123. Westley Weimer, ThanhVu Nguyen, Claire Le Goues, and Stephanie Forrest. Automatically finding patches using genetic programming. In *Proceedings of the 31st International Conference on Software Engineering*, ICSE '09, pages 364–374, Washington, DC, USA, 2009. IEEE Computer Society.

124. Yichen Xie and Alex Aiken. Static detection of security vulnerabilities in scripting languages. In *USENIX-SS'06: Proceedings of the 15th conference on USENIX Security Symposium*, pages 13–13, Berkeley, CA, USA, 2006. USENIX Association.

125. Fang Yu. *Automatic Verification of String Manipulating Programs*. PhD thesis, University of California, Santa Barbara, 2010.

126. Fang Yu, Muath Alkhalaf, and Tevfik Bultan. Generating vulnerability signatures for string manipulating programs using automata-based forward and backward symbolic analyses. In *ASE*, 2009.

127. Fang Yu, Muath Alkhalaf, and Tevfik Bultan. Stranger: An automata-based string analysis tool for php. In *TACAS*, 2010.

128. Fang Yu, Muath Alkhalaf, and Tevfik Bultan. Patching vulnerabilities with sanitization synthesis. In *Proceedings of the 33rd International Conference on Software Engineering (ICSE)*, pages 251–260, 2011.

129. Fang Yu, Muath Alkhalaf, Tevfik Bultan, and Oscar H. Ibarra. Automata-based symbolic string analysis for vulnerability detection. *Formal Methods in System Design*, 44(1):44–70, 2014.

130. Fang Yu, Tevfik Bultan, Marco Cova, and Oscar H. Ibarra. Symbolic string verification: An automata-based approach. In *15th International SPIN Workshop on Model Checking Software (SPIN)*, pages 306–324, 2008.

131. Fang Yu, Tevfik Bultan, and Ben Hardekopf. String abstractions for string verification. In *Proceedings of the 18th International SPIN Conference on Model Checking Software*, pages 20–37, Berlin, Heidelberg, 2011. Springer-Verlag.

132. Fang Yu, Tevfik Bultan, and Oscar H. Ibarra. Symbolic string verification: Combining string analysis and size analysis. In *15th International Conference on Tools and Algorithms for the Construction and Analysis of Systems (TACAS 2009)*, pages 322–336, 2009.

133. Fang Yu, Tevfik Bultan, and Oscar H. Ibarra. Relational string verification using multi-track automata. In *CIAA*, pages 290–299, 2010.

134. Fang Yu, Tevfik Bultan, and Oscar H. Ibarra. Relational string verification using multi-track automata. *Int. J. Found. Comput. Sci.*, 22(8):1909–1924, 2011.

135. Yunhui Zheng, Xiangyu Zhang, and Vijay Ganesh. Z3-str: A z3-based string solver for web application analysis. In *Proceedings of the 9th Joint Meeting on Foundations of Software Engineering (ESEC/FSE)*, pages 114–124, 2013.

136. Edmund M. Clarke, Orna Grumberg, and Doron A. Peled. *Model Checking*. MIT Press, 2001. http://books.google.de/books?id=Nmc4wEaLXFEC

Printed in the United States
By Bookmasters